I0009508

Augmented Reality with Kinect

Develop your own hands-free and attractive
augmented reality applications with Microsoft Kinect

Rui Wang

PUBLISHING

BIRMINGHAM - MUMBAI

Augmented Reality with Kinect

Copyright © 2013 Packt Publishing

All rights reserved. No part of this book may be reproduced, stored in a retrieval system, or transmitted in any form or by any means, without the prior written permission of the publisher, except in the case of brief quotations embedded in critical articles or reviews.

Every effort has been made in the preparation of this book to ensure the accuracy of the information presented. However, the information contained in this book is sold without warranty, either express or implied. Neither the author, nor Packt Publishing, and its dealers and distributors will be held liable for any damages caused or alleged to be caused directly or indirectly by this book.

Packt Publishing has endeavored to provide trademark information about all of the companies and products mentioned in this book by the appropriate use of capitals. However, Packt Publishing cannot guarantee the accuracy of this information.

First published: July 2013

Production Reference: 1040713

Published by Packt Publishing Ltd.
Livery Place
35 Livery Street
Birmingham B3 2PB, UK.

ISBN 978-1-84969-438-4

www.packtpub.com

Cover Image by Suresh Mogre (suresh.mogre.99@gmail.com)

Credits

Author
Rui Wang

Reviewers
Ricard Borràs Navarra

Vangos Pterneas

Acquisition Editor
Kartikey Pandey

Commissioning Editors
Llewellyn Rozario

Priyanka Shah

Technical Editors
Sumedh Patil

Aniruddha Vanage

Copy Editors
Insiya Morbiwala

Alfida Paiva

Laxmi Subramanian

Project Coordinator
Amigya Khurana

Proofreader
Maria Gould

Indexer
Rekha Nair

Graphics
Abhinash Sahu

Production Coordinator
Conidon Miranda

Cover Work
Conidon Miranda

About the Author

Rui Wang is a Software Engineer at Beijing Crystal Digital Technology Co., Ltd. (Crystal CG), in charge of the new media interactive application design and development. He wrote a Chinese book called *OpenSceneGraph Design and Implementation* in 2009. He also wrote the book *OpenSceneGraph 3.0 Beginner's Guide* in 2010 and *OpenSceneGraph 3.0 Cookbook* in 2012, both of which are published by Packt Publishing and co-authored by Xuelei Qian. In his spare time he also writes novels and is a guitar lover.

I must express my deep gratitude to the entire Packt Publishing team for their great work in producing a series of high-quality Mini books, including this one, which introduces the cutting-edge Kinect-based development. Many thanks to Zhao Yang and my other colleagues at Crystal CG, for their talented ideas on some of the recipes in this book. And last but not the least, I'll extend my heartfelt gratitude to my family for their love and support.

About the Reviewers

Ricard Borràs Navarra has always been working around computer vision and machine learning. He started creating machines that apply pattern recognition to quality assortment in the cork production industry. Later, he applied these techniques for human tracking in complex scenarios, creating audience measurement, and people-counter solutions for retail stores.

With the eruption of Kinect, he started working on and deploying augmented reality interactive applications based on this great device. These applications were targeted at the marketing and retail sectors.

All these projects were developed by him as an Inspecta (www.inspecta.es) employee. Also, Ricard has developed several freelance projects based on augmented reality.

Vangos Pterneas is a professional Software Engineer, passionate about natural user interfaces. He has been a Kinect enthusiast ever since the release of the very first unofficial SDKs and has already published a couple of commercial Kinect applications.

He has worked for Microsoft Innovation Center as a .NET developer and consultant and he's now running his own company named LightBuzz. LightBuzz has been awarded the first place in Microsoft's worldwide innovation competition, held in New York.

When he is not coding, he loves blogging about technical stuff and providing the community with open source utilities.

www.PacktPub.com

Support files, eBooks, discount offers and more

You might want to visit www.PacktPub.com for support files and downloads related to your book.

Did you know that Packt offers eBook versions of every book published, with PDF and ePub files available? You can upgrade to the eBook version at www.PacktPub.com and as a print book customer, you are entitled to a discount on the eBook copy. Get in touch with us at service@packtpub.com for more details.

At www.PacktPub.com, you can also read a collection of free technical articles, sign up for a range of free newsletters and receive exclusive discounts and offers on Packt books and eBooks.

http://PacktLib.PacktPub.com

Do you need instant solutions to your IT questions? PacktLib is Packt's online digital book library. Here, you can access, read and search across Packt's entire library of books.

Why Subscribe?

- Fully searchable across every book published by Packt
- Copy and paste, print and bookmark content
- On demand and accessible via web browser

Free Access for Packt account holders

If you have an account with Packt at www.PacktPub.com, you can use this to access PacktLib today and view nine entirely free books. Simply use your login credentials for immediate access.

Table of Contents

Preface

Microsoft Kinect was released in the winter of 2010. As one of the first civil handsfree motion input devices, it brings a lot of fun to end users of Xbox 360 and Windows PCs. And because Kinect is very useful for designing interactive methods in user applications, new media artists and VJs (video jockeys) are also interested in this new technology as it makes their performances more dramatic and mystical.

In this book, we will focus on introducing how to develop C/C++ applications with the Microsoft Kinect SDK, as well as the FreeGLUT library for OpenGL support, and the FreeImage library for image loading. We will cover the topics of Kinect initialization, color and depth image streaming, and skeleton motion and face tracking, and discuss how to implement common gestures with Kinect inputs. A simple but interesting Fruit Ninja-like game will be implemented in the last chapter of this book. Some alternative middlewares and resources will be introduced in the *Appendix*, *Where to Go from Here*, for your reference.

What this book covers

Chapter 1, *Getting Started with Kinect*, shows you how to install Kinect hardware and software on your Windows PC and check if Kinect will start.

Chapter 2, *Creating Your First Program*, demonstrates how to create an OpenGL-based framework first and then initialize the Kinect device in user applications.

Chapter 3, *Rendering the Player*, shows you how to read color and depth images from the Kinect built-in cameras and display them in the OpenGL-based framework. A common way to implement the green screen effect is also discussed.

Chapter 4, *Skeletal Motion and Face Tracking*, demonstrates how to obtain and render the skeleton data calculated by the Kinect sensor. It also introduces the face detection and facial mesh generation APIs with examples.

Chapter 5, *Designing a Touchable User Interface*, shows you how to use Kinect APIs to simulate multi-touch inputs, which are very common in modern interactive applications and GUI developments.

Chapter 6, *Implementing the Scene and Game Play*, demonstrates how to make use of all prior knowledge we have gained to make a Fruit Ninja-like game, which uses Kinect as the input device.

Appendix, *Where to Go from Here*, introduces more alternative middleware and many resource websites for learning and developing Kinect.

What you need for this book

To use this book, you will need a graphics card with robust OpenGL support. It would be better if it is with the latest OpenGL device driver installed from your graphics hardware vendor.

You will also need a working Visual Studio compiler so as to convert C++ source code into executable files. A working Kinect hardware, Microsoft Kinect SDK, and Developer Kit are also required.

Who this book is for

This book is intended for software developers, researchers, and students who are interested in developing Microsoft Kinect-based applications. You should also have basic knowledge of C++ programming before reading this book. Some experience of programming real-time graphics APIs (for example, OpenGL) may be useful, but is not required.

Conventions

In this book, you will find a number of styles of text that distinguish between different kinds of information. Here are some examples of these styles, and an explanation of their meaning.

Code words in text are shown as follows: "The updating of Kinect and user data will be done in the `update()` method."

A block of code is set as follows:

```
#include <GL/freeglut.h>
#include <iostream>

// The updating callback
void update()
{ glutPostRedisplay(); }
```

When we wish to draw your attention to a particular part of a code block, the relevant lines or items are set in bold:

```
#include <GL/freeglut.h>
#include <iostream>

// The updating callback
void update()
{ glutPostRedisplay(); }
```

New terms and **important words** are shown in bold. Words that you see on the screen, in menus or dialog boxes for example, appear in the text like this: "clicking the **Next** button moves you to the next screen".

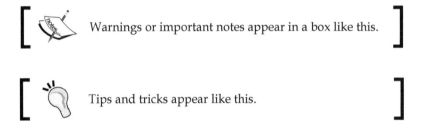

Warnings or important notes appear in a box like this.

Tips and tricks appear like this.

Reader feedback

Feedback from our readers is always welcome. Let us know what you think about this book—what you liked or may have disliked. Reader feedback is important for us to develop titles that you really get the most out of.

To send us general feedback, simply send an e-mail to feedback@packtpub.com, and mention the book title via the subject of your message.

If there is a topic that you have expertise in and you are interested in either writing or contributing to a book, see our author guide on www.packtpub.com/authors.

Customer support

Now that you are the proud owner of a Packt book, we have a number of things to help you to get the most from your purchase.

Downloading the example code

You can download the example code files for all Packt books you have purchased from your account at http://www.packtpub.com. If you purchased this book elsewhere, you can visit http://www.packtpub.com/support and register to have the files e-mailed directly to you.

Errata

Although we have taken every care to ensure the accuracy of our content, mistakes do happen. If you find a mistake in one of our books—maybe a mistake in the text or the code—we would be grateful if you would report this to us. By doing so, you can save other readers from frustration and help us improve subsequent versions of this book. If you find any errata, please report them by visiting http://www.packtpub. com/submit-errata, selecting your book, clicking on the **errata submission form** link, and entering the details of your errata. Once your errata are verified, your submission will be accepted and the errata will be uploaded on our website, or added to any list of existing errata, under the Errata section of that title. Any existing errata can be viewed by selecting your title from http://www.packtpub.com/support.

Piracy

Piracy of copyright material on the Internet is an ongoing problem across all media. At Packt, we take the protection of our copyright and licenses very seriously. If you come across any illegal copies of our works, in any form, on the Internet, please provide us with the location address or website name immediately so that we can pursue a remedy.

Please contact us at copyright@packtpub.com with a link to the suspected pirated material.

We appreciate your help in protecting our authors, and our ability to bring you valuable content.

Questions

You can contact us at questions@packtpub.com if you are having a problem with any aspect of the book, and we will do our best to address it.

1
Getting Started with Kinect

Before the birth of Microsoft Kinect, few people were familiar with the technology of motion sensing. Similar devices have been invented and developed originally for monitoring aerial and undersea aggressors in wars. Then in the non-military cases, motion sensors are widely used in alarm systems, lighting systems and so on, which could detect if someone or something disrupts the waves throughout a room and trigger predefined events. Although radar sensors and modern infrared motion sensors are used more popularly in our life, we seldom notice their existence, and can hardly make use of these devices in our own applications.

But Kinect changed everything from the time it was launched in North America at the end of 2010. Different from most other user input controllers, Kinect enables users to interact with programs without really touching a mouse or a pad, but only through gestures. In a top-level view, a Kinect sensor is made up of an RGB camera, a depth sensor, an IR emitter, and a microphone array, which consists of several microphones for sound and voice recognition. A standard Kinect (for Windows) equipment is shown as follows:

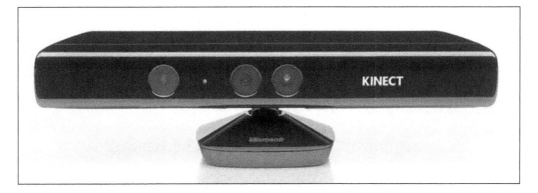

The Kinect device

The Kinect drivers and software, which are either from Microsoft or from third-party companies, can even track and analyze advanced gestures and skeletons of multiple players. All these features make it possible to design brilliant and exciting applications with handsfree user inputs. And until now, Kinect had already brought a lot of games and software to an entirely new level. It is believed to be the bridge between the physical world we exist in and the virtual reality we create, and a completely new way of interacting with arts and a profitable business opportunity for individuals and companies.

In this book, we will try to make an interesting game with the popular Kinect technology for user inputs, with the major components explained gradually in each chapter. As Kinect captures the camera and depth images as video streams, we can also merge this view of our real-world environment with virtual elements, which is called **Augmented Reality (AR)**. This enables users to feel as if they appear and live in a nonexistent world, or something unbelievable exists in the physical earth.

In this chapter, we will first introduce the installation of Kinect hardware and software on personal computers, and then consider a good enough idea compounded of Kinect and augmented reality elements, which will be explained in more detail and implemented in the following chapters.

Before installing the Kinect device on your PCs, obviously you should buy Kinect equipment first. In this book, we will depend on Kinect for Windows or Kinect for Xbox 360, which can be learned about and bought at:

http://www.microsoft.com/en-us/kinectforwindows/

http://www.xbox.com/en-US/kinect

Please note that you don't need to buy an Xbox 360 at all. Kinect will be connected to PCs so that we can make custom programs for it. An alternative choice is Kinect for Windows, which is located at:

http://www.microsoft.com/en-us/kinectforwindows/purchase/

The uses and developments of both will be of no difference for our cases.

Installation of Kinect

It is strongly suggested that you have a *Windows 7 operating system or higher*. It can be either 32-bit or 64-bit and with dual-core or faster processors.

Linux developers can also benefit from third-party drivers and SDKs to manipulate Kinect components, which will be introduced in the *Appendix, Where to Go from Here*, of this book.

Before we start to discuss the software installation, you can download both the Microsoft Kinect SDK and the Developer Toolkit from:

```
http://www.microsoft.com/en-us/kinectforwindows/develop/developer-
downloads.aspx
```

In this book, we prefer to develop Kinect-based applications using Kinect SDK Version 1.5 (or higher versions) and the C++ language. Later versions should be backward compatible so that the source code provided in this book doesn't need to be changed.

Setting up your Kinect software on PCs

After we have downloaded the SDK and the Developer Toolkit, it's time for us to install them on the PC and ensure that they can work with the Kinect hardware. Let's perform the following steps:

1. Run the setup executable with administrator permissions. Select **I agree to the license terms and conditions** after reading the License Agreement.

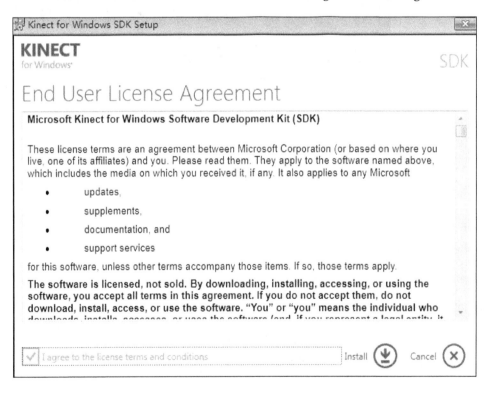

The Kinect SDK setup dialog

2. Follow the steps until the SDK installation has finished. And then, install the toolkit following similar instructions.

3. The hardware installation is easy: plug the ends of the cable into the USB port and a power point, and plug the USB into your PC. Wait for the drivers to be found automatically.

4. Now, start the Developer Toolkit Browser, choose **Samples: C++** from the tabs, and find and run the sample with the name **Skeletal Viewer**.

5. You should be able to see a new window demonstrating the depth/skeleton/color images of the current physical scene, which is similar to the following image:

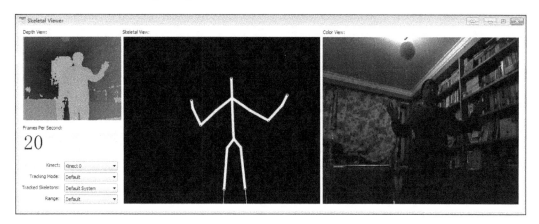

The depth (left), skeleton (middle), and color (right) images read from Kinect

Why did I do that?

We chose to set up the SDK software at first so that it will install the motor and camera drivers, the APIs, and the documentations, as well as the toolkit including resources and samples onto the PC. If the operation steps are inversed, that is, the hardware is connected before installing the SDK, your Windows OS may not be able to recognize the device. Just start the SDK setup at this time and the device should be identified again during the installation process.

But before actually using Kinect, you still have to ensure there is nothing between the device and you (the player). And it's best to keep the play space at least 1.8 m wide and about 1.8 m to 3.6 m long from the sensor. If you have more than one Kinect device, don't keep them face-to-face as there may be infrared interference between them.

If you have multiple Kinects to install on the same PC, please note that one USB root hub can have one and only one Kinect connected. The problem happens because Kinect takes over 50 percent of the USB bandwidth, and it needs an individual USB controller to run. So plugging more than one device on the same USB hub means only one of them will work.

The depth image at the left in the preceding image shows a human (in fact, the author) standing in front of the camera. Some parts may be totally black if they are too near (often less than 80 cm), or too far (often more than 4 m).

 If you are using Kinect for Windows, you can turn on **Near Mode** to show objects that are near the camera; however, Kinect for Xbox 360 doesn't have such features.

You can read more about the software and hardware setup at:

http://www.microsoft.com/en-us/kinectforwindows/purchase/sensor_
setup.aspx

The idea of the AR-based Fruit Ninja game

Now it's time for us to define the goal we are going to achieve in this book. As a quick but practical guide for Kinect and augmented reality, we should be able to make use of the depth detection, video streaming, and motion tracking functionalities in our project. 3D graphics APIs are also important here because virtual elements should also be included and interacted with *irregular* user inputs (not common mouse or keyboard inputs).

A fine example is the Fruit Ninja game, which is already a very popular game all over the world. Especially on mobile devices like smartphones and pads, you can see people destroy different kinds of fruits by touching and swiping their fingers on the screen.

With the help of Kinect, our arms can act as blades to cut off flying fruits, and our images can also be shown along with the virtual environment so that we can determine the posture of our bodies and position of our arms through the screen display.

Unfortunately, this idea is not fresh enough for now. Already, there are commercial products with similar purposes available in the market; for example:

```
http://marketplace.xbox.com/en-US/Product/Fruit-Ninja-
Kinect/66acd000-77fe-1000-9115-d80258410b79
```

But please note that we are not going to design a completely different product here, or even bring it to the market after finishing this book. We will only learn how to develop Kinect-based applications, work in our own way from the very beginning, and benefit from the experience in our professional work or as an amateur. So it is okay to reinvent the wheel this time, and have fun in the process and the results.

Summary

Kinect, which is a portmanteau of the words "kinetic" and "connect", is a motion sensor developed and released by Microsoft. It provides a **natural user interface (NUI)** for tracking and manipulating handsfree user inputs such as gestures and skeleton motions. It can be considered as one of the most successful consumer electronics device in recent years, and we will be using this novel device to build the Fruit Ninja game in this book.

We will focus on developing Kinect and AR-based applications on Windows 7 or higher using the Microsoft Kinect SDK 1.5 (or higher) and the C++ programming language. Mainly, we have introduced how to install Kinect for Windows SDK in this chapter. Linux and Mac OS X users can first read the *Appendix*, *Where to Go from Here*, which provides an alternative method to call Kinect functionalities on other systems. Developers of .NET, processing, or other language tools may also find useful resources in the last chapter of this book.

In the next chapter, we will learn how to start and shut down the Kinect device in our applications, and prepare a basic framework for further development.

2
Creating Your First Program

We have already introduced how to `install` the Kinect device on Windows in the previous chapter, as well as some official examples showing the basic concepts of Kinect programming. In this chapter, we will prepare a simple OpenGL framework for our Kinect-based game using the C++ language. OpenGL is a well-rounded and evolving cross-platform API for rendering 2D and 3D graphics. It supports multiple languages including C/C++, Java, Python, and C#. As we are working on an Augmented Reality (AR) project, which must consist of the view of the real world and some virtual elements, OpenGL will be a good choice here because of its hardware-accelerated features and popularity all over the world.

 As you may know, Microsoft's DirectX is another reliable 2D/3D graphics API that could fit our requirements. But it is only used under Windows currently, and can hardly support languages except C/C++ and C#. You can learn more about DirectX at:

http://msdn.microsoft.com/en-us/library/ee663274.aspx

Also, we have discussed about the installation of Kinect for the Windows SDK in the previous chapter. The SDK provides us libraries and header files for use in the official toolkits and our own applications. To make use of the SDK, we will have to include the header files and link to the necessary libraries in our projects to generate the final executables.

To note, this is not a Kinect API reference book, so we can't list and introduce all the functions and structures here. You can refer to the following website for more information about the Kinect NUI API:

http://msdn.microsoft.com/en-us/library/hh855366.aspx

Before starting to create a C++ project, you should at least have a C++ compiler. For Windows users, the Visual Studio product is always a better choice. You can download Visual C++ 2012 Express for free from:

`http://www.microsoft.com/visualstudio/eng/downloads#d-2012-express`

It is assumed that you are already familiar with this environment, as well as the C++ language and OpenGL programming. You can refer to the following link for details about the OpenGL API and related resources at `http://www.opengl.org/`.

Preparing the development environment

Before we create a new console project in Visual Studio, we are first going to download some third-party dependencies for convenience; otherwise we will have to implement the construction of the OpenGL context completely by ourselves, and some important but complex functionality such as image reading and displaying. To make this process less lengthy, we will choose only two external libraries here:

* FreeGLUT
* FreeImage

Both of them are open source and easy to understand.

External libraries will be used as dependencies for each of our C++ projects, so we can make use of the OpenGL-based drawing and image loading functions, which are important for future work.

FreeGLUT is a complete alternative to the famous **OpenGL Utility Toolkit (GLUT)**, allowing developers to create and manage OpenGL contexts with just a few functions and readable callbacks. Its official website is:

`http://freeglut.sourceforge.net/`

But, we can directly download the prebuilt Windows package containing the DLL, import library, and header files from:

`http://files.transmissionzero.co.uk/software/development/GLUT/`
`freeglut-MSVC.zip`

Then unzip the downloaded file to a proper path for later use.

FreeImage is a fast and stable library for reading and writing to many popular graphics image formats. It also provides basic image manipulations such as rotating, resizing, and color adjustment. Its official website is:

`http://freeimage.sourceforge.net/`

The prebuilt Windows package is located at:

`http://downloads.sourceforge.net/freeimage/FreeImage3153Win32.zip`

Download and unzip it too. Find the DLLs, library files, and headers of both, and place them in separate subdirectories (for example, `bin`, `lib`, and `include`), so that we can manage and use these dependencies efficiently, as shown in the following screenshot:

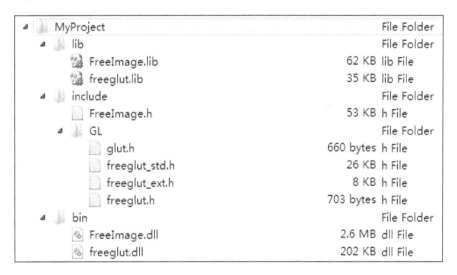

An example of the directory structure of dependencies

Other dependencies include OpenGL and the Kinect SDK. The OpenGL library is automatically integrated with every Visual Studio project, so we don't have to worry about the installation.

From now on, the variable `${MYPROJECT_DIR}` will be used to indicate the project folder, which contains both project files and all third-party dependencies, and `${KINECTSDK10_DIR}` is used to indicate the location of the Microsoft Kinect SDK. You can either set these two environment variables in the Windows system, or replace it manually with the actual paths while setting project properties.

Building the Visual Studio project

Now, we are going to create our first application with Visual Studio. Please note that the Microsoft Kinect SDK is only for Windows users. If you want to develop Kinect-based applications on other platforms, you may prefer OpenNI instead, which is also introduced in the previous chapter of this book.

1. Create a new C++ console project by navigating to **File | New | Project** and choose **Win32 Console Application** from the **Visual C++** menu. Set the project name to FirstProgram or any name you like. Select **Empty project** in the next dialog and click on **OK**.

2. Add **include directories** by right-clicking on the project and selecting **Properties**. Navigate to **C/C++ | General | Additional Include Directories** and input the following dependency paths:

    ```
    ${MYPROJECT_DIR}/include; ${KINECTSDK10_DIR}/inc
    ```

3. Navigate to **Linker | Input | Additional Dependencies** in the Property page. Input the library paths and names as follows:

    ```
    opengl32.lib; glu32.lib; ${MYPROJECT_DIR}/lib/freeglut.lib;
    ${MYPROJECT_DIR}/lib/FreeImage.lib;
    ${KINECTSDK10_DIR}/lib/x86/Kinect10.lib
    ```

     For X64 configuration, we should use ${KINECTSDK10_DIR}/lib/amd64/Kinect10.lib instead, and make sure the FreeGLUT and FreeImage libraries are also built for X64 systems (you may have to build them from the source code).

4. Now add some initial code to make our first program work.

    ```cpp
    #include <GL/freeglut.h>
    #include <iostream>

    // The updating callback
    void update()
    { glutPostRedisplay(); }

    // The rendering callback
    void render()
    { glutSwapBuffers(); }

    // The window resizing callback
    void reshape( int w, int h )
    { glViewport( 0, 0, w, h ); }
    ```

```
// The keyboard callback: make sure we can exit when press
// Esc or 'Q' key.
void keyEvents( unsigned char key, int x, int y )
{
    switch ( key )
    {
    case 27: case 'Q': case 'q':
        glutLeaveMainLoop();
        return;
    }
    glutPostRedisplay();
}

int main( int argc, char** argv )
{
    // Initialize a GLUT window and make it full-screen
    glutInit( &argc, argv );
    glutInitDisplayMode(
        GLUT_RGB|GLUT_DOUBLE|GLUT_DEPTH|GLUT_MULTISAMPLE );
    glutCreateWindow( "ch2_01_OpenGL_Env" );
    glutFullScreen();

    // Register necessary callbacks
    glutIdleFunc( update );
    glutDisplayFunc( render );
    glutReshapeFunc( reshape );
    glutKeyboardFunc( keyEvents );

    // Start the main loop
    glutMainLoop();
    return 0;
}
```

Downloading the example code

You can download the example code files for all Packt books you have purchased from your account at http://www.packtpub.com. If you purchased this book elsewhere, you can visit http://www.packtpub.com/support and register to have the files e-mailed directly to you.

5. The source code may be a little long for beginners to read. But this is exactly the structure we are going to use in the entire book. The updating of Kinect and user data will be done in update(). The rendering of color/depth images and the skeleton, as well as other virtual objects will be done in render().

6. It's still a little too complicated and too redundant to write all the source code for loading textures with FreeImage, or drawing meshes with OpenGL vertex array features. Fortunately, we have already provided some useful functions for immediate use.

7. Drag the files in the common folder (from the downloaded package) onto your project icon. Now you should have two more source files named GLUtilities.cpp and TextureManager.cpp.

8. Compile and build all the source code into an executable file. Put the output into ${MYPROJECT_DIR}/bin and run it. You will only see a black screen, which we can quit by pressing the *Esc* key.

Here, we just create a standard console application with FreeGLUT, FreeImage, and Kinect SDK as dependencies. The framework we created cannot do anything at present but we will soon add something interesting to make it more colorful.

Please note that the added source files GLUtilities.cpp and TextureManager.cpp are not used. But they will play an important role in the following examples to render all kinds of textures and geometries.

> If you are still not familiar enough with the OpenGL API, or want to do something interesting before stepping into Kinect programming, there are some good references for you to read and test at:
>
> - The OpenGL official website: http://www.opengl.org/
> - The OpenGL wiki page: http://en.wikipedia.org/wiki/OpenGL
> - The NeHe OpenGL examples: http://nehe.gamedev.net/ (good for exercising; see the Legacy Tutorials section)

Starting the device

Now it's time to initialize the Kinect device in our own application. There will be a lot of Kinect API functions for us to use without any preparatory lessons. But don't worry; you will find that most of them are self-explanatory and easy to understand. Also, we will introduce each function and their parameters in the *Understanding the code* section.

We will continue working on the framework we have just created, so existing code lines will not be listed here again.

Initializing and using Kinect in C++

Now we can try to find and start the Kinect device in our own C++ framework.

1. Add the following include files:

```
#include <MSHTML.h>
#include <NuiApi.h>
#include <sstream>
```

2. Add the necessary global variables for use in all functions:

```
INuiSensor* context = NULL;
HANDLE colorStreamHandle = NULL;
HANDLE depthStreamHandle = NULL;
std::string hudText;
```

3. Add an `initializeKinect()` function, which will be called before the GLUT main loop. It returns `false` if the process fails for any reason.

```
// Check if there are any Kinect sensors connected with
// current PC and obtain the number
int numKinects = 0;
HRESULT hr = NuiGetSensorCount( &numKinects );
if ( FAILED(hr) || numKinects<=0 ) return false;

// Create the sensor object and set it to context.
// Here we only use the first device (index 0) we find.
hr = NuiCreateSensorByIndex( 0, &context );
if ( FAILED(hr) ) return false;

// Initialize the sensor with color/depth/skeleton enabled
DWORD nuiFlags = NUI_INITIALIZE_FLAG_USES_SKELETON |
                 NUI_INITIALIZE_FLAG_USES_COLOR |
                 NUI_INITIALIZE_FLAG_USES_DEPTH;
hr = context->NuiInitialize( nuiFlags );
if ( FAILED(hr) ) return false;

// Open color and depth video streams for capturing.
// The resolution is set to 640x480 here.
hr = context->NuiImageStreamOpen(
    NUI_IMAGE_TYPE_COLOR, NUI_IMAGE_RESOLUTION_640x480,
    0, 2, NULL, &colorStreamHandle );
if ( FAILED(hr) ) return false;
```

```
hr = context->NuiImageStreamOpen(
    NUI_IMAGE_TYPE_DEPTH, NUI_IMAGE_RESOLUTION_640x480,
    0, 2, NULL, &depthStreamHandle );
if ( FAILED(hr) ) return false;

// Enable skeleton tracking
hr = context->NuiSkeletonTrackingEnable( NULL, 0 );
if ( FAILED(hr) ) return false;
return true;
```

4. Add a `destroyKinect()` function after the main loop in which we just release the sensor object we created before.

```
if ( context )
    context->NuiShutdown();
```

5. In the main entry, we alter the last few lines as follows:

```
if ( !initializeKinect() ) return 1;
glutMainLoop();
destroyKinect();
return 0;
```

6. The program can compile and run now. But it still produces nothing. We don't know whether Kinect works or not as it shows a blank window. So, next we will add a few lines in `update()` and `render()` to print some continuous updated Kinect information.

7. At the beginning of `update()`, obtain one color frame and output the current frame number and time stamp value into a string:

```
NUI_IMAGE_FRAME colorFrame;
HRESULT hr = context->NuiImageStreamGetNextFrame(
    colorStreamHandle, 0, &colorFrame );
if ( SUCCEEDED(hr) )
{
    std::stringstream ss;
    ss << "Frame: " << colorFrame.dwFrameNumber << "   "
        << "Time: " <<
(double)colorFrame.liTimeStamp.QuadPart * 0.001;
    hudText = ss.str();
    context->NuiImageStreamReleaseFrame(
        colorStreamHandle, &colorFrame );
}
```

8. In the `render()` function, render the text on screen as follows:

```
// Clear last frame buffer
glClearColor( 0.0f, 0.0f, 0.0f, 0.0f );
glClear( GL_COLOR_BUFFER_BIT|GL_DEPTH_BUFFER_BIT );

// Set up the projection matrix for text display
glMatrixMode( GL_PROJECTION );
glLoadIdentity();
glOrtho( 0.0, 1.0, 0.0, 1.0, -1.0, 1.0 );

glMatrixMode( GL_MODELVIEW );
glLoadIdentity();

// Print the text at the bottom of the window
glRasterPos2f( 0.01f, 0.01f );
glColor4f( 1.0f, 1.0f, 1.0f, 1.0f );
glutBitmapString( GLUT_BITMAP_TIMES_ROMAN_24,
                  (const unsigned char*)hudText.c_str() );
```

9. Now, execute the compiled program; you may find it a little slower when you start executing. Be patient until the depth sensor starts to lighten. The screen is still dark but you will find a line of animation text at the bottom-left as shown in the following figure:

A snapshot of the application

10. If the application directly exits without displaying anything, the initialization process may fail. Add some text before returning false to see the value of `hr` in `initializeKinect()`; also check if your Kinect sensor is connected and not used by other programs.

Understanding the code

The following table shows all the Kinect functions we have used as well as descriptions of the important parameters.

Function/method name	Parameters	Description
NuiGetSensorCount	int* pCount	Get the number of Kinect sensors connected to the PC and set to pCount.
NuiCreateSensorByIndex	int index, INuiSensor** ppNuiSensor	Creates an instance of the Kinect sensor at index and sets it to ppNuiSensor.
INuiSensor::NuiInitialize	DWORD dwFlags	Initialize Kinect with specified feature options, including audio, color, depth, depth with player index, and skeleton.
INuiSensor::NuiImageStreamOpen	NUI_IMAGE_TYPE type, NUI_IMAGE_RESOLUTION res, DWORD dwFrameFlags, DWORD dwFrameLimit, HANDLE hNextFrameEvent, HANDLE *phStreamHandle	• Open an image stream with specific type (color, depth, and so on) and resolution (640x480 for our case), and set its handle to phStreamHandle. • The dwFrameFlags provides some additional options (default is 0). • The dwFrameLimit means frame numbers limited for buffering (always set to 2). • The hNextFrameEvent is used for multithreaded cases.
INuiSensor::NuiSkeletonTrackingEnable	HANDLE hNextFrameEvent, DWORD dwFlags	Enable skeleton tracking with additional options dwFlags (default is 0). The hNextFrameEvent is used for multithreaded cases.
INuiSensor::NuiShutdown		Turns Kinect off.

 All functions and methods start with the prefix "Nui". It is just short for **Natural User Interface(NUI)**.

Now the total process of creating and using Kinect in user applications can be summarized as follows:

1. Find and create the sensor object.

2. Initialize the sensor with the required features (image streams and skeleton tracking), and enable these features.

3. Update every frame to get stream and skeleton data for use.

4. Release the sensor object when exiting.

Quite simple, isn't it? Note that, we didn't introduce the lines in the `update()` function here. We will explain that in the next chapter, with more interesting live images shown on the screen instead of a boring line of text.

Additional information

Another interesting and challenging task is to implement a multithreaded version of the initialization and updating of Kinect. In fact, some functions here have already supported such uses by accepting event handles as parameters, including `NuiImageStreamOpen()` and `NuiSkeletonTrackingEnable()`. Events will change when new video stream/skeleton frames arrive, so we can listen to them with `WaitForMultipleObjects()` in a separate thread and then obtain related frame data.

Summary

In this chapter, we have created a simple enough OpenGL framework to use it throughout the whole book. It is built on the open source FreeGLUT and FreeImage libraries, as well as the Microsoft Kinect SDK.

We have also successfully initialized the Kinect device in our own C++ application and obtained some continuous updating values for displaying on screen, which can be found in the downloadable source code package of this book. This means that Kinect is now ready for extensive usage based on color/depth camera images and skeleton tracking points. These will be covered in the next two chapters.

3
Rendering the Player

In the previous chapter, we successfully initialized the Kinect device and started the color and depth image streams. Now it is time to read data from the streams at every frame and show them on screen so that we can obtain the very important depth values for many uses. For example, we can change the color image at a specified depth to a different one. Another good idea is to place some virtual objects around the player while rendering the scene. Then we use depths to decide if part of an object is behind or in front of the player to make the compositing result more natural.

In the following sections, we will introduce how to obtain and render color images read by the cameras and depth images read by the depth sensor. We will continue to work on the framework created in the previous chapter, with Kinect already initialized and ready for use.

Choosing image stream types

The Microsoft Kinect SDK supports several types of image streams. In the example from the previous chapter, we started the device with the following options:

```
DWORD nuiFlags = NUI_INITIALIZE_FLAG_USES_SKELETON |
    NUI_INITIALIZE_FLAG_USES_COLOR |
    NUI_INITIALIZE_FLAG_USES_DEPTH;
hr = context->NuiInitialize( nuiFlags );
```

It can be concluded from the enumeration values that three types of data will be allocated in the current application: the skeleton, the color, and the depth.

The color and depth data are of course necessary here because we are going to make use of them soon. The skeleton data will be first used in *Chapter 4, Skeletal Motion and Face Tracking*. Besides this, Kinect even provides two more types of data for us to select at the beginning of the application:

- `NUI_INITIALIZE_FLAG_USES_AUDIO`: This provides the audio data from the microphone array located at the bottom of the device.

- `NUI_INITIALIZE_FLAG_USES_DEPTH_AND_PLAYER_INDEX`: This provides the depth data with each pixel marked with a player index. This data type will be widely used in our book to quickly separate the player image from the background.

To open the stream for reading, we call `NuiImageStreamOpen()` immediately after the device initialization. It contains a resolution parameter as follows:

```
context->NuiImageStreamOpen(
    NUI_IMAGE_TYPE_COLOR, NUI_IMAGE_RESOLUTION_640x480,
    0, 2, NULL, &colorStreamHandle );
```

The `NUI_IMAGE_RESOLUTION_640x480` value tells us that the color data is allocated as a 640 x 480 sized image from the camera. Other available values include `NUI_IMAGE_RESOLUTION_80x60`, `NUI_IMAGE_RESOLUTION_320x240`, and `NUI_IMAGE_RESOLUTION_1280x960`. Only the resolution of the color stream can be set to 1280 x 960. Depth data can't have such high resolutions at present, but it is always enough to use a resolution of 320 x 240 or 640 x 480 for depth.

Obtaining color and depth images

In this section, we are going to first learn how to obtain and display images from Kinect on the screen. OpenGL uses textures to contain image data and maps textures onto any mesh surface so that the images can be shown as surface colors. The mapping process requires the texture coordinates of surfaces, known as UVW coordinates in most 3D modeling software, to denote the texture's x/y/z axes for correct image painting. A good tutorial about the OpenGL texture mapping can be found at `http://www.glprogramming.com/red/chapter09.html`.

As we are not developing very complicated 3D applications, we can just simplify the whole process into two steps:

- Creating a simple quad with four vertices that is always facing the screen, and mapping a 2D texture (with only UV coordinates set) onto it for displaying 2D images
- Generating and updating the texture from Kinect image data every frame

Now let's start.

Drawing color and depth as textures

We will first obtain color and depth data from the Kinect device and render both onto the OpenGL quad to show them on the screen. The color and depth images should be updated and assigned to the OpenGL texture per frame:

1. We will first add some necessary headers and global variables for use based on the previous example. The `TextureObject` structure is defined in `GLUtilities.h` for storing the attributes of an OpenGL texture:

```
#include "common/TextureManager.h"
#include "common/GLUtilities.h"
TextureObject* colorTexture = NULL;
TextureObject* depthTexture = NULL;
```

2. These two texture objects can be created in the main entry before starting the simulation loop, and deleted after it:

```
colorTexture = createTexture(640, 480, GL_RGB, 3);
depthTexture = createTexture(640, 480, GL_LUMINANCE, 1);

glutMainLoop();

destroyTexture( colorTexture );
destroyTexture( depthTexture );
```

3. Now we are going to update the color/depth streams and obtain usable data from every frame's image. The method to acquire one frame was already used in the previous chapter, but we will explain its usage in more detail in the next section:

```
// Update the color image
NUI_IMAGE_FRAME colorFrame;
HRESULT hr = context->NuiImageStreamGetNextFrame(
    colorStreamHandle, 0, &colorFrame );
if ( SUCCEEDED(hr) )
{
    updateImageFrame( colorFrame, false );
    context->NuiImageStreamReleaseFrame (
        colorStreamHandle, &colorFrame );
}
```

```
// Update the depth image
NUI_IMAGE_FRAME depthFrame;
hr = context->NuiImageStreamGetNextFrame(
    depthStreamHandle, 0, &depthFrame );
if ( SUCCEEDED(hr) )
{
    updateImageFrame( depthFrame, true );
    context->NuiImageStreamReleaseFrame (
        depthStreamHandle, &depthFrame );
}
```

4. A new updateImageFrame() function is introduced here. It includes two parameters: imageFrame, containing the real data, and isDepthFrame to differentiate between the two kinds of streams:

```
void updateImageFrame (
    NUI_IMAGE_FRAME& imageFrame, bool isDepthFrame )
{
    ...
}
```

5. The content of the new function is shown in the following code snippet. It demonstrates how to load the image data into the corresponding OpenGL texture object:

```
INuiFrameTexture* nuiTexture = imageFrame.pFrameTexture;
NUI_LOCKED_RECT lockedRect;
nuiTexture->LockRect( 0, &lockedRect, NULL, 0 );
if ( lockedRect.Pitch!=NULL )
{
    // We assume the image is always 640 x 480 as set in
    // the initialization process.
    const BYTE* buffer = (const BYTE*)lockedRect.pBits;
    for ( int i=0; i<480; ++i )
    {
        const BYTE* line = buffer + i * lockedRect.Pitch;
        const USHORT* bufferWord = (const USHORT*)line;
        for ( int j=0; j<640; ++j )
        {
            if ( !isDepthFrame )
            {
                // For colors, convert each pixel from BGR
                // to RGB
                unsigned char* ptr = colorTexture->
                    bits + 3 * (i * 640 + j);
```

```
                    *(ptr + 0) = line[4 * j + 2];
                    *(ptr + 1) = line[4 * j + 1];
                    *(ptr + 2) = line[4 * j + 0];
                }
                else
                {
                    // For depth, extract the depth value part
                    unsigned char* ptr = depthTexture->
                        bits + (i * 640 + j);
                    *ptr = (unsigned char)
                        NuiDepthPixelToDepth(bufferWord[j]);
                }
            }
        }
    }

    // Send the textures to OpenGL side for displaying
    TextureObject* tobj =
        (isDepthFrame ? depthTexture : colorTexture);
    glBindTexture( GL_TEXTURE_2D, tobj->id );
    glTexImage2D( GL_TEXTURE_2D, 0, tobj->internalFormat,
        tobj->width, tobj->height, 0, tobj->imageFormat,
        GL_UNSIGNED_BYTE, tobj->bits );
}
nuiTexture->UnlockRect( 0 );
```

6. Now in the `render()` function, we render the color and depth images horizontally on the screen at the same time. This can be easily done with the functions provided in `GLUtilities.h`, with both vertices and texture coordinates set:

```
// Define vertices and texture coordinates for a simple
// quad
// The quad's width is half of the screen's width so it can
// be drawn twice to show two different images on the
// screen
GLfloat vertices[][3] = {
    { 0.0f, 0.0f, 0.0f }, { 0.5f, 0.0f, 0.0f },
    { 0.5f, 1.0f, 0.0f }, { 0.0f, 1.0f, 0.0f }
};
GLfloat texcoords[][2] = {
    {0.0f, 1.0f}, {1.0f, 1.0f}, {1.0f, 0.0f}, {0.0f, 0.0f}
};
VertexData meshData = { &(vertices[0][0]), NULL, NULL,
    &(texcoords[0][0]) };
```

```
// Draw the quad with color texture attached
glBindTexture( GL_TEXTURE_2D, colorTexture->id );
drawSimpleMesh( WITH_POSITION|WITH_TEXCOORD, 4,
    meshData, GL_QUADS );

// Change position and draw the quad with depth texture
glTranslatef( 0.5f, 0.0f, 0.0f );
glBindTexture( GL_TEXTURE_2D, depthTexture->id );
drawSimpleMesh( WITH_POSITION|WITH_TEXCOORD, 4,
    meshData, GL_QUADS );
```

7. Compile and run the application. You will see the screen is split into two parts to show both the color and depth images:

The stream outputs of Kinect

8. You will find that the depth image (right) represents the depth value of each pixel in the color image (left). Objects near the depth sensor will turn out to be darker. When out of the range of the depth sensor (80-90 cm for Kinect on Xbox and 40-50 cm for Kinect on Windows), the depth will be painted black and the value becomes unsuitable for practical use.

> The color image resolution might be higher than the depth image resolution, so one depth pixel may correspond to more than one color pixel. But in this example, we set the same resolution for both the images.

Understanding the code

Before we obtain image data from the stream, we should request the frame data using `NuiImageStreamGetNextFrame()`, which accepts three parameters: the handle, the time to wait in milliseconds before returning (0 ms in our case), and the `NUI_IMAGE_FRAME` pointer, which will contain real image data. If we choose not to wait before obtaining every frame's data, this function may fail sometimes because of being called too frequently; so, we must check the return value at every frame to see if the function has succeeded, and then call `NuiImageStreamReleaseFrame()` to release the data at last.

The `NUI_IMAGE_FRAME` structure has a member `pFrameTexture` variable containing the image resource we need (in the previous chapter, we made use of `dwFrameNumber` and `liTimeStamp` from it, which indicate the frame number and timestamp of the most recent frame). But before directly reading data from it, we should first lock the buffer so that it is only available for the current thread to use and unlock the buffer after receiving the data. This is done in `updateImageFrame()` using a `NUI_LOCKED_RECT` variable:

```
INuiFrameTexture* nuiTexture = imageFrame.pFrameTexture;
NUI_LOCKED_RECT lockedRect;
nuiTexture->LockRect( 0, &lockedRect, NULL, 0 );

nuiTexture->UnlockRect( 0 );
```

Locking the buffer before working on it is important because Kinect may transfer new data to the buffer and at the same time, other threads, or even other applications, may also handle the same data. Any operations on the buffer without locking it are totally unsafe and prohibited by the Kinect SDK.

After locking the buffer, we can now safely use the `NUI_LOCKED_RECT` structure to read frame image data. Its definition is as follows:

```
typedef struct {
    INT Pitch;      // The number of bytes of data in a row
    int size;       // The size of pBits, in bytes
    BYTE *pBits;    // A pointer to the upper left corner of data
                    // rect
} NUI_LOCKED_RECT;
```

The pBits starts from the upper-left corner, but OpenGL treats image data assuming that it starts from the lower-left corner. So the result rendered on screen will be flipped vertically if we don't care about the difference. In this example, we use OpenGL texture coordinates to solve the problem, as you can see from the following diagram:

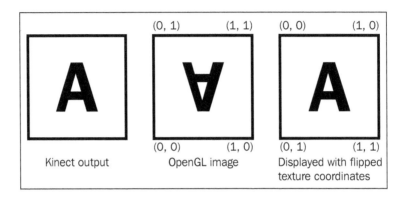

The Kinect image displayed with OpenGL; because the OpenGL image is already flipped after being read, the texture coordinates are set to flip the image again on the quad

The functions used in this example are listed in the following table:

Function/method name	Parameters	Description
INuiSensor::NuiImageStreamGetNextFrame	HANDLE stream, DWORD timeToWait, and const NUI_IMAGE_FRAME **ppcImageFrame	This function gets the next frame of data from the specified image stream and sets it to ppcImageFrame, waiting for timeToWait milliseconds.
INuiSensor::NuiImageStreamReleaseFrame	HANDLE stream and const NUI_IMAGE_FRAME *pImageFrame	This function releases the data frame pImageFrame from the specified handle stream.
INuiFrameTexture::LockRect	UINT Level, NUI_LOCKED_RECT *pLockedRect, RECT *pRect, and DWORD Flags	This function locks the buffer and sets it to pLockedRect. Other parameters are unused.

Function/method name	Parameters	Description
`INuiFrameTexture::UnlockRect`	`UINT Level`	This function unlocks the buffer. The parameter here is unused and must be set to `0`.

An incorrect way to combine depth and color

The basic idea of a green screen, or chroma key compositing, is commonly used in the film industry. The director shoots a video with a single-colored backdrop (always green or blue), and then replaces the single colors with another video or still image. This produces some exciting effects such as the actor running out of an explosion field or the weather broadcaster standing in front of a large virtual map/earth. In this chapter, we will try to implement the same effect with the Kinect device.

The Kinect device is designed to be able to resolve depth data from the sensor to human body results. It can recognize both the entire body and different parts of human limbs and tries placing the joints to build up a skeleton, which is perhaps the most impressive feature of Kinect. In fact, we had never seen a for-civil-use production before that can perform similar work.

To learn more about the skeleton recognition of Kinect, you can refer to `http://research.microsoft.com/en-us/projects/vrkinect/default.aspx`.

But in this example, it is enough to only know where the human body is in the depth image. For the depth pixels within a specific human body, Kinect will save the player index (a non-zero number). So the only task for us is to read the player indices in the depth image and clear the pixels in the color image if their corresponding depth value doesn't have a player index.

A traditional way for background subtraction

The simplest idea to combine depth and color images is to display them at the same position but allow color pixels to be shown only when the depth value at the same row and column is valid. We will implement our example in this way and see if it works:

1. The depth stream we used before only recorded values read from the depth sensor. The Kinect SDK also provides a packed depth stream with both depth and player index recorded in every pixel. This is very useful for our case.

2. First, we modify the `initializeKinect()` function to listen to the packed depth stream instead of the original one. The only change here is to replace `NUI_INITIALIZE_FLAG_USES_DEPTH` with `NUI_INITIALIZE_FLAG_USES_DEPTH_AND_PLAYER_INDEX`:

```
DWORD nuiFlags = NUI_INITIALIZE_FLAG_USES_SKELETON |
    NUI_INITIALIZE_FLAG_USES_COLOR |
    NUI_INITIALIZE_FLAG_USES_DEPTH_AND_PLAYER_INDEX;
hr = context->NuiInitialize( nuiFlags );


hr = context->NuiImageStreamOpen(
    NUI_IMAGE_TYPE_DEPTH_AND_PLAYER_INDEX,
    NUI_IMAGE_RESOLUTION_640x480,
    0, 2, NULL, &depthStreamHandle );
```

3. The second step seems straightforward. Because we know which pixel contains a valid player index and which does not, we can just set the corresponding texture's pixel to 0 where no player index is found. The `depthTexture` then works like a mask image, replacing the same locations in the color texture with empty values:

```
unsigned char* ptr = depthTexture->bits + (i * 640 + j);
if ( NuiDepthPixelToPlayerIndex(bufferWord[j])>0 )
    *ptr = 255;
else
    *ptr = 0;
```

4. OpenGL can implement masking with the `GL_BLEND` feature, as shown in the following code snippet:

```
// Define vertices and texture coordinates for a simple
// quad
// The quad will cover whole screen to show the final image
GLfloat vertices[][3] = {
    { 0.0f, 0.0f, 0.0f }, { 1.0f, 0.0f, 0.0f },
    { 1.0f, 1.0f, 0.0f }, { 0.0f, 1.0f, 0.0f }
};
GLfloat texcoords[][2] = {
    {0.0f, 1.0f}, {1.0f, 1.0f}, {1.0f, 0.0f}, {0.0f, 0.0f}
};
VertexData meshData =
    { &(vertices[0][0]), NULL, NULL, &(texcoords[0][0]) };

// Draw the quad with color texture attached
glBindTexture( GL_TEXTURE_2D, colorTexture->id );
drawSimpleMesh( WITH_POSITION|WITH_TEXCOORD, 4,
    meshData, GL_QUADS );
```

```
// Enable blending with the depth texture color as factors
glEnable( GL_BLEND );
glBlendFunc( GL_ONE_MINUS_SRC_COLOR, GL_SRC_COLOR );

// Draw the quad again before the previous one and blend
// them
// Result will be the product of color and depth textures
glTranslatef( 0.0f, 0.0f, 0.1f );
glBindTexture( GL_TEXTURE_2D, depthTexture->id );
drawSimpleMesh
    ( WITH_POSITION|WITH_TEXCOORD, 4, meshData, GL_QUADS );

// Disable blending at last
glDisable( GL_BLEND );
```

5. So the result should be nice, shouldn't it? Let's compile and stand before the Kinect device to see if background subtraction has been successfully implemented:

The result of our background subtraction, which is not good at all

6. Maybe you will be disappointed now. The depth data is clipped correctly, but it is not aligned with the color image at all! The player is obviously slanted and thus makes the entire application unusable.

Understanding the code

The NUI_IMAGE_TYPE_DEPTH_AND_PLAYER_INDEX data type is a slightly different stream from the depth sensor. It doesn't contain pure depth value, but one that is combined with a 3-bit index value belonging to specific players determined by the skeleton tracking system. This packed depth pixel will thus have 11 bits and must be stored using a USHORT data type. The Kinect SDK provides two convenient functions to read the real depth value and player index from every pixel read: NuiDepthPixelToDepth() and NuiDepthPixelToPlayerIndex().

In this example, we decide the value of our depth texture using the player index so that we can get an image with only white (with the player index) and black (without the player index) pixels. In OpenGL, we blend this monochrome picture with the color texture. White pixels are transparent so the colors become visible, and black pixels are still black so background colors appear blank, thus generating the final image. Unfortunately, this is incorrect.

The reason is simple. Kinect's depth and color images in fact come from different sensors. They may have different **fields of view** (**FOV**) and not face the same direction. So a pixel at a specified location in a depth image is not always at the same location in a color image. Without considering these factors, we can hardly line up the depth and image pixels and produce a correct green screen effect.

However, thanks to the Kinect SDK, we still have some methods to fix this problem, such as mapping a pixel at a certain location in depth space to the corresponding coordinates in color space. We could even directly use some functions to achieve this.

Aligning color with depth

The steps to implement the green screen have now changed because of the alignment problem of color and depth images. Instead of directly blending the depth and color images, we will first construct a new texture for storing remapped colors (and use the player index to subtract the background colors). Then we will display the new texture on screen, which can be treated as the result of background removal.

Generating a color image from depth

This time we will use an inbuilt Kinect API method to align the color data with depth and combine them again. Let's start now:

1. Now we will have to traverse all pixels and save the color values where the player index is valid (for others, we set the color to total black). This requires a new texture object, which is named `playerColorTexture` here:

   ```
   TextureObject* colorTexture = NULL;
   TextureObject* playerColorTexture = NULL;
   ```

2. The `playerColorTexture` is in RGBA format, so we can use its alpha channel for image masking, besides the RGB components for normal color display:

   ```
   colorTexture = createTexture(640, 480, GL_RGB, 3);
   playerColorTexture = createTexture(640, 480, GL_RGBA, 4);

   glutMainLoop();

   destroyTexture( colorTexture );
   destroyTexture( playerColorTexture );
   ```

3. In the `updateImageFrame()` function, we will try to read and compute the player color texture instead of the depth one. This will be done in a new `setPlayerColorPixel()` function:

   ```
   if ( !isDepthFrame )
   {
       unsigned char* ptr = colorTexture->bits + 3 *
           (i * 640 + j);
       *(ptr + 0) = line[4 * j + 2];
       *(ptr + 1) = line[4 * j + 1];
       *(ptr + 2) = line[4 * j + 0];
   }
   else
       setPlayerColorPixel( bufferWord[j], j, i );
   ```

4. The `setPlayerColorPixel()` function has three parameters: the `depthValue` read from Kinect and the x and y values located in the image space. It returns `false` if the current location doesn't have a player index attached:

   ```
   bool setPlayerColorPixel
       ( const USHORT depthValue, int x, int y );
   ```

5. The content of the function is listed here:

```
// Find correct place to write the RGBA value
unsigned char* ptr =
    playerColorTexture->bits + 4 * (y * 640 + x);

// Check if there exists a player index
if ( NuiDepthPixelToPlayerIndex(depthValue)>0 )
{
    // Get correct x and y coordinates in color image space
    LONG colorX = 0, colorY = 0;
    context->NuiImageGetColorPixelCoordinates
        FromDepthPixelAtResolution(
        NUI_IMAGE_RESOLUTION_640x480,
        NUI_IMAGE_RESOLUTION_640x480, NULL,
        x, y, depthValue, &colorX, &colorY );
    if ( colorX>=640 || colorY>=480 ) return false;

    // Write color value to the playerColorTexture
    unsigned char* colorPtr = colorTexture->bits + 3 *
        (colorY * 640 + colorX);
    *(ptr + 0) = *(colorPtr + 0);
    *(ptr + 1) = *(colorPtr + 1);
    *(ptr + 2) = *(colorPtr + 2);
    *(ptr + 3) = 255;
}
else
{
    // Write 0 to all four components of playerColorTexture
    *(ptr + 0) = 0;
    *(ptr + 1) = 0;
    *(ptr + 2) = 0;
    *(ptr + 3) = 0;
}
return true;
```

6. We will render the image quad again with only the player color texture this time:

```
// Define vertices and texture coordinates for a simple
// quad
// The quad will cover whole screen
GLfloat vertices[][3] = {
    { 0.0f, 0.0f, 0.0f }, { 1.0f, 0.0f, 0.0f },
    { 1.0f, 1.0f, 0.0f }, { 0.0f, 1.0f, 0.0f }
};
GLfloat texcoords[][2] = {
    {0.0f, 1.0f}, {1.0f, 1.0f}, {1.0f, 0.0f}, {0.0f, 0.0f}
};
VertexData meshData =
    { &(vertices[0][0]), NULL, NULL, &(texcoords[0][0]) };

// Render the player's color image
glBindTexture( GL_TEXTURE_2D, playerColorTexture->id );
drawSimpleMesh( WITH_POSITION|WITH_TEXCOORD, 4,
    meshData, GL_QUADS );
```

7. Now let's see if there are any improvements compared to the previous one:

The result of background subtraction, which is acceptable

8. The result is acceptable now. The aliasing problem occuring at the edges of the depth image is still distinct. We can use some image-based methods to optimize it later, but for this chapter, it is enough!

Understanding the code

The `NuiImageGetColorPixelCoordinatesFromDepthPixelAtResolution()` function is the key for this recipe as it converts depth space coordinates to color space, lining up pixels in both the images correctly. This function has quite a few parameters:

```
HRESULT
    NuiImageGetColorPixelCoordinatesFromDepthPixelAtResolution(
    NUI_IMAGE_RESOLUTION eColorResolution,
    NUI_IMAGE_RESOLUTION eDepthResolution,
    const NUI_IMAGE_VIEW_AREA *pcViewArea,
    LONG lDepthX,
    LONG lDepthY,
    USHORT usDepthValue,
    LONG *plColorX,
    LONG *plColorY
);
```

Here `eColorResolution` and `eDepthResolution` are set to `NUI_IMAGE_RESOLUTION_640x480` because we initialize both the streams with the same settings. The `pcViewArea` parameter is `NULL` here because we don't need any optional zoom and pan settings for the color image. The `lDepthX` and `lDepthY` parameters are x/y coordinates in depth image space and `usDepthValue` is the depth value. With these inputs, Kinect can compute the coordinate offset and output coordinates in color space to `plColorX` and `plColorY`.

The given depth coordinates should not be too near to the depth image bounds, otherwise they may be mapped to coordinates outside the bounds of the color image, which is invalid and should be checked and excluded.

> The `usDepthValue` parameter must be the original depth value, that is, the packed pixel value for `NUI_IMAGE_TYPE_DEPTH_AND_PLAYER_INDEX`, not the extracted one from `NuiDepthPixelToDepth()`.

Additional information

The `NuiImageGetColorPixelCoordinatesFromDepthPixelAtResolution()` function is not effective because it must convert the pixels one by one. The Kinect SDK also provides another function called `NuiImageGetColorPixelCoordinateFrameFromDepthPixelFrameAtResolution()` to perform the same work on an array of depth values, and the output will be an array of color coordinate values:

```
HRESULT NuiImageGetColorPixelCoordinateFrameFrom
    DepthPixelFrameAtResolution(
    NUI_IMAGE_RESOLUTION eColorResolution,
    NUI_IMAGE_RESOLUTION eDepthResolution,
    DWORD cDepthValues,
    USHORT* pDepthValues,
    DWORD cColorCoordinates,
    LONG* pColorCoordinates
);
```

Can you try to use it to modify this example and make it work smoothly?
Note that pColorCoordinates is an array of x/y coordinates for each pixel.

Using a green screen with Kinect

Now we can develop a simple game to satisfy ourselves, which will also be used
as part of our Fruit Ninja game. The idea can be described as a magic photographer
who automatically puts the photo of the player in front of the Kinect device onto any
scenery images, pretending that he had taken this photo some time ago.

The example we just finished is used to show a player with a single colored
background, so the only work left is to load a still image from the disk and blend
it with the player image to produce a final composite photo.

Making a magic photographer

Let's continue working on the previous example code we created, which already
contains the kernel functionality for our use:

1. We need a background image to be shown under the player's image. The
   alpha channel of the player texture will be used to decide if the background
   should show or not. FreeImage is used to load the image from a disk file and
   bind it to an OpenGL texture. The global ID for the texture is declared here:

   ```
   const unsigned int backgroundTexID = 1;
   ```

2. In the main entry, we will read a file named background.bmp from the disk.
   Please copy any of your image files to the executable directory and convert
   it to BMP format for use. Note that FreeImage always loads images in BGR
   format; that is, in blue-green-red order:

   ```
   if ( TextureManager::Inst()->LoadTexture
       ("background.bmp", backgroundTexID, GL_BGR_EXT) )
   {
   ```

```
        glTexParameteri
        ( GL_TEXTURE_2D, GL_TEXTURE_MIN_FILTER, GL_LINEAR );
        glTexParameteri
        ( GL_TEXTURE_2D, GL_TEXTURE_MAG_FILTER, GL_LINEAR );
}
```

3. Render simple OpenGL quads with Kinect color and depth images:

```
// Define vertices and texture coordinates for a simple
// quad
// The quad will cover whole screen to show the final image
GLfloat vertices[][3] = {
    { 0.0f, 0.0f, 0.0f }, { 1.0f, 0.0f, 0.0f },
    { 1.0f, 1.0f, 0.0f }, { 0.0f, 1.0f, 0.0f }
};
GLfloat texcoords[][2] = {
    {0.0f, 1.0f}, {1.0f, 1.0f}, {1.0f, 0.0f}, {0.0f, 0.0f}
};
VertexData meshData =
    { &(vertices[0][0]), NULL, NULL, &(texcoords[0][0]) };

// Draw the background image quad
TextureManager::Inst()->BindTexture( backgroundTexID );
drawSimpleMesh
    ( WITH_POSITION|WITH_TEXCOORD, 4, meshData, GL_QUADS );

// Enable blending with player texture alpha as factors
glEnable( GL_BLEND );
glBlendFunc( GL_SRC_ALPHA, GL_ONE_MINUS_SRC_ALPHA );

// Draw the quad again before the previous one and blend
// them
glTranslatef( 0.0f, 0.0f, 0.1f );
glBindTexture( GL_TEXTURE_2D, playerColorTexture->id );
drawSimpleMesh
    ( WITH_POSITION|WITH_TEXCOORD, 4, meshData, GL_QUADS );
```

4. Start the program and you will see your chosen scenery image shown on the screen, which could be your favorite place:

A sand landscape chosen by the author

5. Stand in front of the Kinect device and you will be added to the scene now. Find a good position for yourself and take a photo now:

Add the player into the scene

Understanding the code

The only difference between this recipe and the previous one is that a background image is added and blended with the player. The alpha values we set in the previous recipe play an important role because they are used as the OpenGL blending factor as follows:

```
glBlendFunc( GL_SRC_ALPHA, GL_ONE_MINUS_SRC_ALPHA );
```

This means the source pixels S, which form the player image to be drawn, and the target pixels T, which are the background colors, will be blended using the following equation:

```
S * alpha + T * (1 - alpha)
```

So, the player pixels will only be rendered on screen when alpha is 1, and the background pixels are kept where alpha is 0.

 Although we actually set alpha to 0 or 255 in the program, it is always mapped to [0, 1] in OpenGL for further use.

Additional information

The composite image is still not that good because of aliasing and flickering at the player edges. One possible improvement is to blur the depth image before using it. We could also do some postprocessing work on the generated image to perfectly match it with the background. It is now up to you to consider implementing these features using, for instance, GrabCut (http://research.microsoft.com/en-us/um/cambridge/projects/visionimagevideoediting/segmentation/grabcut.htm).

Summary

In this chapter, we saw how to obtain color and depth images from the Kinect sensors and display them in the OpenGL context. A special type of depth image, which packs both depth value and player index in one pixel, was also introduced and used for background subtraction. It is more powerful than the traditional green screen technique because it doesn't need a single colored background (but the precision for civil-level use is not too high).

The main problem we encountered here was that the depth and color pixels were not aligned. The Kinect SDK also provides some functions for quickly mapping depth space coordinates to color space ones so that we can combine these two streams smoothly for different uses.

4
Skeletal Motion and Face Tracking

Capturing and tracking skeleton images of one or two people is one of the most exciting features of Kinect development. It can transform many ideas to reality, including gesture recognition, multi-touch emulation, data-driven character animations, and even some advanced techniques such as motion capture and model reconstruction. The skeletal mapping work in every Kinect device is actually done by a microprocessor in the sensor (or directly by the Xbox core), and the results can be retrieved using corresponding APIs for use in our own applications.

The Microsoft Kinect SDK 1.5 also includes a new face tracking module that can track the position and rotation of our heads, and the shapes of our eyes and mouth. It even provides APIs to compute a virtual face mesh, which can be directly rendered in the 3D world. We will also introduce these excellent functionalities in this chapter, although they are not quite related to our planned Fruit Ninja game.

The face tracker API may not be located in the same directory of the Kinect SDK. If you have already installed the Developer Toolkit as discussed in *Chapter 1, Getting Started with Kinect*, you should be able to locate it at ${FTSDK_DIR}. Here, the environment variable indicates the location of the Kinect Developer Toolkit.

Understanding skeletal mapping

At present, Microsoft Kinect can identify up to six people within the view of the field, but it can only track at most two people in detail at the same time.

The players must stand (or sit) in front of the Kinect device, facing the sensors. If the player shows only a part of his body to the sensors, or wants the sensors to recognize sideways poses, the result may not be accurate, as some part of the skeleton may be in the wrong place, or may jitter back and forth.

Usually, the player is suggested to stand between 0.8 m and 4.0 m away from the device. Kinect for Windows may perform better for near distances because it has a near depth range mode (0.4 m) for use.

In every frame, Kinect will calculate a skeleton image for each person in tracking, which includes 20 joints to represent a complete human body. The positions and meanings of these joints can be found in the following figure:

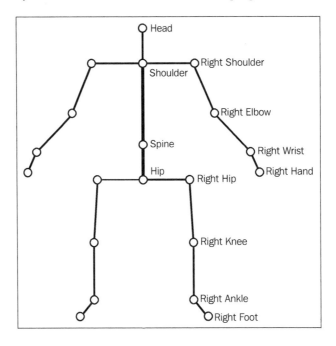

The skeleton mapping

 Kinect uses infrared lights to calculate the depth of people and reconstructs the skeleton accordingly. So if you are using multiple Kinect devices for more precise skeleton mapping or other purposes, a different infrared light source (including another Kinect) in the view of the field will interfere with the current device and thus reduce the precision of computation. The interference may be low but we still have to avoid such a problem in practical situations.

Obtaining joint positions

Before we can consider using the skeleton for gesture-based interaction, we should first print out all the skeletal-joint-related data to have a directly perceived look of the Kinect skeleton positions. The data can then be merged with our color image so that we can see how they are matched with each other in real time.

Drawing the skeleton

We will first draw the skeleton with a series of lines to see how Kinect defines all the skeletal bones.

1. The Microsoft Kinect SDK uses `NUI_SKELETON_POSITION_COUNT` (equivalent to 20 for the current SDK version) to represent the number of joints of one skeleton, so we define an array to store their positions.

   ```
   GLfloat skeletonVertices[NUI_SKELETON_POSITION_COUNT][3];
   ```

2. Add the following lines for updating a skeleton frame in the `update()` function.

   ```
   NUI_SKELETON_FRAME skeletonFrame = {0};
   hr = context->NuiSkeletonGetNextFrame( 0, &skeletonFrame );
   if ( SUCCEEDED(hr) )
   {
       // Traverse all possible skeletons in tracking
       for ( int n=0; n<NUI_SKELETON_COUNT; ++n )
       {
           // Check each skeleton data to see if it is tracked
           NUI_SKELETON_DATA&
   data=skeletonFrame.SkeletonData[n];
           if ( data.eTrackingState==NUI_SKELETON_TRACKED )
           {
               updateSkeletonData( data );
               break;  // in this demo, only handle one skeleton
           }
       }
   }
   ```

3. We declare a new function named `updateSkeletonData()` with one
    `NUI_SKELETON_DATA` argument for handling the specified skeleton data.
    Now let's fill it.

```
POINT coordInDepth;
USHORT depth = 0;

// Traverse all joints
for ( int i=0; i<NUI_SKELETON_POSITION_COUNT; ++i )
{
    // Obtain joint position and transform it to depth
space
    NuiTransformSkeletonToDepthImage(
        data.SkeletonPositions[i],
        &coordInDepth.x, &coordInDepth.y,
        &depth, NUI_IMAGE_RESOLUTION_640x480 );

    // Transform all coordinates to [0, 1] and set them
    // to the array we defined before.
    // We will discuss about the transformation later
    skeletonVertices[i][0] =
        (GLfloat)coordInDepth.x / 640.0f;
    skeletonVertices[i][1] =
        1.0f - (GLfloat)coordInDepth.y / 480.0f;
    skeletonVertices[i][2] =
        (GLfloat)NuiDepthPixelToDepth(depth) * 0.00025f;
}
```

4. Before rendering the skeleton data we retrieved in `updateSkeletonData()`,
    we have to define the skeleton index array so that OpenGL knows how
    to connect these joint points. Because we will only draw the skeleton as
    a reference, it is sufficient to draw the points using the `GL_LINES` mode,
    which indicates that every two points are connected to form a line segment.

5. Using the human skeleton figure we just saw, we can quickly write out the
    definition as follows:

```
// Every two indices will form a line-segment
// All Kinect enumerations here should be self-explained
GLuint skeletonIndices[38] = {
NUI_SKELETON_POSITION_HIP_CENTER,
NUI_SKELETON_POSITION_SPINE,
```

```
    NUI_SKELETON_POSITION_SPINE,
    NUI_SKELETON_POSITION_SHOULDER_CENTER,
    NUI_SKELETON_POSITION_SHOULDER_CENTER,
    NUI_SKELETON_POSITION_HEAD,
    // Left arm
    NUI_SKELETON_POSITION_SHOULDER_LEFT,
    NUI_SKELETON_POSITION_ELBOW_LEFT,
    NUI_SKELETON_POSITION_ELBOW_LEFT,
    NUI_SKELETON_POSITION_WRIST_LEFT,
    NUI_SKELETON_POSITION_WRIST_LEFT,
    NUI_SKELETON_POSITION_HAND_LEFT,
    // Right arm
    NUI_SKELETON_POSITION_SHOULDER_RIGHT,
    NUI_SKELETON_POSITION_ELBOW_RIGHT,
    NUI_SKELETON_POSITION_ELBOW_RIGHT,
    NUI_SKELETON_POSITION_WRIST_RIGHT,
    NUI_SKELETON_POSITION_WRIST_RIGHT,
    NUI_SKELETON_POSITION_HAND_RIGHT,
    // Left leg
    NUI_SKELETON_POSITION_HIP_LEFT,
    NUI_SKELETON_POSITION_KNEE_LEFT,
    NUI_SKELETON_POSITION_KNEE_LEFT,
    NUI_SKELETON_POSITION_ANKLE_LEFT,
    NUI_SKELETON_POSITION_ANKLE_LEFT,
    NUI_SKELETON_POSITION_FOOT_LEFT,
    // Right leg
    NUI_SKELETON_POSITION_HIP_RIGHT,
    NUI_SKELETON_POSITION_KNEE_RIGHT,
    NUI_SKELETON_POSITION_KNEE_RIGHT,
    NUI_SKELETON_POSITION_ANKLE_RIGHT,
        NUI_SKELETON_POSITION_ANKLE_RIGHT,
    NUI_SKELETON_POSITION_FOOT_RIGHT,
    // Others
    NUI_SKELETON_POSITION_SHOULDER_CENTER,
    NUI_SKELETON_POSITION_SHOULDER_LEFT,
    NUI_SKELETON_POSITION_SHOULDER_CENTER,
    NUI_SKELETON_POSITION_SHOULDER_RIGHT,
    NUI_SKELETON_POSITION_HIP_CENTER,
    NUI_SKELETON_POSITION_HIP_LEFT,
    NUI_SKELETON_POSITION_HIP_CENTER,
    NUI_SKELETON_POSITION_HIP_RIGHT
    };
```

6. Now in the `render()` function, we render the skeleton lines along with the color image. The `drawIndexedMesh()` function is declared in `GLUtilities.h` for quick and convenient use.

```
glDisable( GL_TEXTURE_2D );
glLineWidth( 5.0f );

VertexData skeletonData = { &(skeletonVertices[0][0]),
NULL, NULL, NULL };
drawIndexedMesh( WITH_POSITION,
NUI_SKELETON_POSITION_COUNT, skeletonData, GL_LINES, 38,
skeletonIndices );
```

7. Compile the program, run it, and see what we have just done!

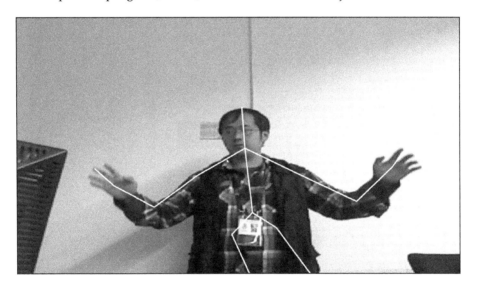

The skeleton along with the color image (only the upper body)

8. Please note that the author has only shown the upper body in the Kinect camera, so the lower body part of skeletal mapping result may be incorrect and might shake. But the shoulder and arms perform very well.

Understanding the code

Some new functions we used in this example are listed as follows:

Function/method name	Parameters	Description
NuiSkeletonGetNextFrame	DWORD timeToWait and NUI_SKELETON_ FRAME* frame	Gets skeleton data of the current frame and sets it to the frame structure.
NuiTransformSkeletonToDepthImage	Vector4 point, LONG* coordX, LONG* coordY, USHORT* depth and NUI_IMAGE_ RESOLUTION res	Returns the depth space coordX, coordY, and depth with specified resolution res and specified point in skeleton space.

In this example, we will transform the joint positions to depth space, and then to [0, 1] using the following functions:

```
NuiTransformSkeletonToDepthImage(
        data.SkeletonPositions[i],
        &coordInDepth.x, &coordInDepth.y,
        &depth, NUI_IMAGE_RESOLUTION_640x480 );
skeletonVertices[i][0] = (GLfloat)coordInDepth.x / 640.0f;
skeletonVertices[i][1] = 1.0f - (GLfloat)coordInDepth.y / 480.0f;
skeletonVertices[i][2] = (GLfloat)NuiDepthPixelToDepth(depth) *
0.00025f;
```

The original skeleton positions are stored in `data.SkeletonPositions` to define all the necessary joints in world space. The Microsoft Kinect SDK uses a right-handed coordinate system with values in meters to manage all the positions. The origin is at the sensor pinhole, and the z axis is pointing from the sensor to the view field. So when we lift our hands horizontally (x axis), the coordinates of our left and right hands may be:

```
Left hand: (-0.8, 0.2, 2.0), and right hand: (0.8, 0.2, 2.0)
```

The actual values won't be so stable and symmetrical, but it indicates that the person is standing about 2 meters away from the Kinect device, and his/her arm span is about 1.6 meters.

It can be easily inferred from the code segment that, after mapping the positions to depth space, every joint's x and y values are using image coordinates (640 x 480), and the z value is the actual depth, which is equivalent to the depth image pixel at the same location. As we know, Kinect can detect depth values from 0.8 meters to at most 4 meters, so we divide the return value of `NuiDepthPixelToDepth()` with 4000 millimeters to re-project the joint's z axis to [0, 1], which assumes that z = 0 is the camera lens plane (but can't reach it), and z = 1 is the farthest.

The Microsoft Kinect SDK also provides joint orientation data, which could be useful if we want to control the virtual body more precisely. Local or world space orientations can be obtained by calling the following function in the `updateSkeletonData()` function:

```
NUI_SKELETON_BONE_ORIENTATION
rotations[NUI_SKELETON_POSITION_COUNT];
NuiSkeletonCalculateBoneOrientations( &data, rotations );
```

It will output to an array containing the orientation data of each joint. Try to get and display it (for example, add a small axis at each joint position) on screen by yourself.

Drawing the linetrails following the hands

Now let's start to develop a very important part of our Fruit Ninja game: the knives that cut any coming fruits. Our hands can simulate the knives very well here, because in a motion-sense environment, they are always the most agile and accurate objects to operate on anything in space.

It will be easy to know the per-frame positions of the two hands as shown in the previous example. But it is also a good idea to add some trailing effects to demonstrate how fast and sharp the knives are, and to indicate to the players where their weapons are. In this example, we will emulate these trails with a series of continuous line segments. The alpha values of each line segment can also change so that the entire trail seems to fade out at the end.

Drawing the path for specified joints

To implement linetrails of two hands, we have to use a dynamic array to store historical points that the hands have moved to. We then connect them to implement the line-trail effect, as shown in the following steps:

1. We directly define two STL vector elements to store a custom structure to simplify the process. Besides, a fixed-size array is used to store the fading colors of the point list.

```
struct Vertex { GLfloat x, y, z; };
std::vector<Vertex> leftHandTrails;
std::vector<Vertex> rightHandTrails;
GLfloat trailColors[20][4];

// Use this to create a translucent depth texture
TextureObject* playerDepthTexture = NULL;
```

2. In the main entry, we set the trailing color array to constant values. The alpha channel values will change to express a fade-in effect (index 0 is nearly invisible, and index 19 is opaque).

```
for ( int i=0; i<20; ++i )
{
    trailColors[i][0] = 1.0f;
    trailColors[i][1] = 1.0f;
    trailColors[i][2] = 1.0f;
    trailColors[i][3] = (float)(i + 1) / 20.0f;
}

// The optimized depth texture will have two channels, one
for player index display and another for transparency.
playerDepthTexture = createTexture(640, 480,
GL_LUMINANCE_ALPHA, 2);
```

3. Obtain the skeleton data and set the coordinates to the hand trailing vectors.

```
POINT coordInDepth;
USHORT depth = 0;
for ( int i=0; i<NUI_SKELETON_POSITION_COUNT; ++i )
{
    NuiTransformSkeletonToDepthImage(
        data.SkeletonPositions[i], &coordInDepth.x,
&coordInDepth.y,
        &depth, NUI_IMAGE_RESOLUTION_640x480 );

    Vertex vertex;
    vertex.x = (GLfloat)coordInDepth.x;
    vertex.y = 1.0f - (GLfloat)coordInDepth.y / 480.0f;
    vertex.z = (GLfloat)NuiDepthPixelToDepth(depth) *
0.00025f;
```

```
if ( i==NUI_SKELETON_POSITION_HAND_LEFT )
{
    // Add latest hand point to the vector
    // Remove the oldest one if the vector is too large
    leftHandTrails.push_back( vertex );
    if ( leftHandTrails.size()>20 ) leftHandTrails.erase(
leftHandTrails.begin() );
}
else if ( i==NUI_SKELETON_POSITION_HAND_RIGHT )
{
    // Do the same thing as handling the left hand
    rightHandTrails.push_back( vertex );
    if ( rightHandTrails.size()>20 ) rightHandTrails.erase(
rightHandTrails.begin() );
}
}
```

4. The `updateImageFrame()` function should also be changed to write to the new `playerDepthTexture` object. If there's a player, a translucent white pixel is written; otherwise a totally transparent pixel is written.

```
const BYTE* buffer = (const BYTE*)lockedRect.pBits;
for ( int i=0; i<480; ++i )
{
    const BYTE* line = buffer + i * lockedRect.Pitch;
    const USHORT* bufferWord = (const USHORT*)line;
    for ( int j=0; j<640; ++j )
    {
        unsigned char* ptr = playerDepthTexture->bits + 2 *
(i * 640 + j);
        if ( NuiDepthPixelToPlayerIndex(bufferWord[j])>0 )
        {
            *(ptr + 0) = 200;
            *(ptr + 1) = 80;
        }
        else
        {
            *(ptr + 0) = 0;
            *(ptr + 1) = 0;
        }
    }
}
```

5. Enable blending and draw the depth texture in `render()`. Before doing this, we should first place a background image as the shown in the previous chapter, so that the result will be a smooth light shadow added onto a scenery image.

```
glEnable( GL_BLEND );
glBlendFunc( GL_SRC_ALPHA, GL_ONE_MINUS_SRC_ALPHA );
glTranslatef( 0.0f, 0.0f, 0.1f );

glBindTexture( GL_TEXTURE_2D, playerDepthTexture->id );
drawSimpleMesh( WITH_POSITION|WITH_TEXCOORD, 4, meshData,
GL_QUADS );
```

6. At last, we draw the hand linetrails as line strips. Don't disable blending before the hand trails are finished, so that the alpha channel of `trailColors` will take effect.

```
glDisable( GL_TEXTURE_2D );
glLineWidth( 50.0f );

VertexData leftHandData = { &(leftHandTrails[0].x), NULL,
&(trailColors[0][0]), NULL };
drawSimpleMesh( WITH_POSITION|WITH_COLOR,
leftHandTrails.size(), leftHandData, GL_LINE_STRIP );

VertexData rightHandData = { &(rightHandTrails[0].x), NULL,
&(trailColors[0][0]), NULL };
drawSimpleMesh( WITH_POSITION|WITH_COLOR,
rightHandTrails.size(), rightHandData, GL_LINE_STRIP );

// Disable blending so it won't affect the next frame
glDisable( GL_BLEND );
```

7. Now let's have a look at the final result:

The linetrails effect with a translucent depth

Understanding the code

This example does not provide us with more technical details, but instead provides us with some more ideas. The skeleton data from Kinect can be used to locate some important body parts in an accurate and efficient way.

Thus, the skeletal mapping feature may help to quickly add virtual objects onto or around the body. For example, add a hat on the head, a jackboot on the foot, butterflies around the player, and so on.

One big problem is that this joint-related data may jitter all the time if the environment is not good enough or the player cannot show his full body in front of the sensor. A very common improvement is to record some historical data and compute the average values of the current data and old data, as shown in this example. Another solution for smooth skeleton transformation will be discussed again in the next chapter.

Face tracking in Kinect

Face detecting and tracking is a famous computer vision technology. It analyzes the images from a webcam or other input devices, and tries to determine the locations and sizes of human faces from these inputs. Some detailed face parts can also be guessed from the given images, including eyes, eyebrows, nose, and the mouth. We can even determine the emotion of a specific face, or the identity of a human from the face tracking results.

The Microsoft Kinect SDK supports face tracking from Version 1.5 onwards. It requires color and depth images from the sensors (or customized sources) as inputs, and returns the position data of the detected head, as well as some important recording points on the face, all of which can be retrieved or used for reconstructing the 3D face mesh in real time.

We will explain how the Microsoft Kinect SDK declares and generates the face mesh at the end of this chapter.

Detecting a face from the camera

Before we start to detect a face in front of the Kinect camera, we should set up the include and library files of the face tracking library. For Visual Studio users, first add `${FTSDK_DIR}/inc` to the current include directories of your project, and add `${FTSDK_DIR}/Lib/x86/FaceTrackLib.lib` to the additional dependencies. You may also have to manually copy all the files in `${FTSDK_DIR}/Redist/x86` to your executable folder to help find these dynamic library files smoothly.

Now we will work on a Kinect example we have done before, which has already initialized the Kinect context for the new face tracking library that is ready to use.

Detecting and drawing the face rectangle

In this example, we will first detect the position and size of the face.

1. Include the face tracking library header, and declare some global variables for recording necessary face tracking data.

   ```
   #include <FaceTrackLib.h>

   IFTFaceTracker* tracker = NULL;
   IFTResult* faceResult = NULL;
   FT_SENSOR_DATA sensorData;
   RECT faceRect;
   bool isFaceTracked = false;
   ```

2. The face tracking process requires both the color and depth buffers for computation, so we define them here, too.

   ```
   TextureObject* colorTexture = NULL;
   TextureObject* packedDepthTexture = NULL;
   ```

3. Create a new `initializeFaceTracker()` function to initialize the library.

   ```
   bool initializeFaceTracker()
   {
       // Create the face tracker object
       tracker = FTCreateFaceTracker();
       if ( !tracker )
       {
           std::cout << "Can't create face tracker" <<
   std::endl;
           return false;
       }

       // Define parameters for both color and depth sensors.
       FT_CAMERA_CONFIG colorConfig = {640, 480,
   NUI_CAMERA_COLOR_NOMINAL_FOCAL_LENGTH_IN_PIXELS};
       FT_CAMERA_CONFIG depthConfig = {640, 480,
   NUI_CAMERA_DEPTH_NOMINAL_FOCAL_LENGTH_IN_PIXELS * 2};

       // Initialize the face tracker with those parameters
       HRESULT hr = tracker->Initialize( &colorConfig,
   &depthConfig, NULL, NULL );
   ```

```
    if ( FAILED(hr) )
    {
        std::cout << "Can't initialize face tracker" <<
std::endl;
        return false;
    }

    // Create the face tracker result object
    hr = tracker->CreateFTResult( &faceResult );
    if ( FAILED(hr) )
    {
        std::cout << "Can't create face tracker result" <<
std::endl;
        return false;
    }

    // Create sensor frames for storing color and depth
data
    sensorData.pVideoFrame = FTCreateImage();
    sensorData.pDepthFrame = FTCreateImage();
    if ( !sensorData.pDepthFrame || !sensorData.pDepthFrame
)
    {
        std::cout << "Can't create color/depth images" <<
std::endl;
        return false;
    }

    // Attach the texture object data (which is updated by
    // Kinect NUI API) to the face tracker sensor frames,
    // to actually connect sensor data and the face tracker
    sensorData.pVideoFrame->Attach( 640, 480,
(void*)colorTexture->bits, FTIMAGEFORMAT_UINT8_R8G8B8,
640*3 );
    sensorData.pDepthFrame->Attach( 640, 480,
(void*)packedDepthTexture->bits,
FTIMAGEFORMAT_UINT16_D13P3, 640 );

    // Other default values for the sensor data
    sensorData.ZoomFactor = 1.0f;
    sensorData.ViewOffset.x = 0;
    sensorData.ViewOffset.y = 0;
    return true;
}
```

4. Use `destroyFaceTracker()` to release face objects safely.

```
bool destroyFaceTracker()
{
    if ( faceResult ) faceResult->Release();
    if ( tracker ) tracker->Release();
    return true;
}
```

5. In the main entry, we will successively initialize the Kinect device, the color and depth texture, and the face tracker. The releasing process is inverse, so the face tracker function is destroyed first and the Kinect at last.

```
if ( !initializeKinect() ) return 1;
colorTexture = createTexture(640, 480, GL_RGB, 3);
packedDepthTexture = createTexture(640, 480,
GL_LUMINANCE_ALPHA, 2);
if ( !initializeFaceTracker() ) return 1;

glutMainLoop();

destroyFaceTracker();
destroyTexture( colorTexture );
destroyTexture( packedDepthTexture );
destroyKinect();
```

6. The `updateImageFrame()` function will write the BGR colors to the color texture as usual, but will write the entire packed depth values to the depth texture. This depth texture will not be used for displaying purposes, but will be used for the face tracker to infer the result.

```
const BYTE* buffer = (const BYTE*)lockedRect.pBits;
for ( int i=0; i<480; ++i )
{
    const BYTE* line = buffer + i * lockedRect.Pitch;
    const USHORT* bufferWord = (const USHORT*)line;
    for ( int j=0; j<640; ++j )
    {
        if ( !isDepthFrame )
        {
            unsigned char* ptr = colorTexture->bits + 3 *
(i * 640 + j);
            *(ptr + 0) = line[4 * j + 2];
            *(ptr + 1) = line[4 * j + 1];
            *(ptr + 2) = line[4 * j + 0];
        }
```

```
        else
        {
            USHORT* ptr = (USHORT*)packedDepthTexture->bits
+ (i * 640 + j);
            *ptr = bufferWord[j];
        }
    }
}
```

7. In the `update()` function, we will have a simple choice: if no faces are found in the last frame, we restart the tracking work using the sensor data we defined before; if there was one face in tracking, we continue the tracking process to see if the face is still distinguishable.

```
if ( !isFaceTracked )
{
    // Start new tracking process if no face is in scope
    hr = tracker->StartTracking( &sensorData, NULL, NULL,
faceResult );
    if ( SUCCEEDED(hr) && SUCCEEDED(faceResult-
>GetStatus()) ) isFaceTracked = true;
}
else
{
    // Continue the tracking process if a face is already
    // captured in the field of view
    hr = tracker->ContinueTracking( &sensorData, NULL,
faceResult );
    if ( FAILED(hr) || FAILED(faceResult->GetStatus()) )
isFaceTracked = false;
}
```

8. Now we are going to draw a 2D quad around the tracked face in the `render()` function. The result is stored in the `faceRect` variable and displayed in every frame.

```
// Re-obtain the face rectangle when there are tracked
faces
if ( isFaceTracked )
    faceResult->GetFaceRect( &faceRect );
```

```
// The calculated rectangle is in image coordinates,
// so re-project it to [0, 1] for rendering in OpenGL
float l = (float)faceRect.left / 640.0f;
float r = (float)faceRect.right / 640.0f;
float b = 1.0f - (float)faceRect.bottom / 480.0f;
float t = 1.0f - (float)faceRect.top / 480.0f;
GLfloat faceVertices[][3] = {
    { l, b, 0.1f }, { r, b, 0.1f }, { r, t, 0.1f }, { l, t,
0.1f }
};
VertexData faceData = { &(faceVertices[0][0]), NULL, NULL,
NULL };

// Draw looped line-segments around the face
glDisable( GL_TEXTURE_2D );
glLineWidth( 5.0f );
drawSimpleMesh( WITH_POSITION, 4, faceData, GL_LINE_LOOP );
```

9. Assuming that we have already rendered the color image as the background image, the rendering result may be similar to the following screenshot:

The face tracking quad

Understanding the code

The functions and methods used in this example are listed in the following table:

Function/method name	Parameters	Description
`FTCreateFaceTracker`		Creates an `IFTFaceTracker` face tracker instance.
`FTCreateImage`		Creates an `IFTImage` image-object instance.

`IFTFaceTracker::Initialize`	`const FT_CAMERA_CONFIG* color,`
	
	`const FT_CAMERA_CONFIG* depth,`
	
	`FTRegisterDepthToColor func,`
	
	`PCWSTR modelPath`
		Initializes the face tracker with suitable camera configurations for `color` and `depth` sensors. The optional `func` can be used to register a function for depth-to-color mapping, and the optional `modelPath` is the path of an external face model.

`IFTFaceTracker::CreateFTResult`	`IFTResult** ppResult`	Creates an `IFTResult` result object instance.

`IFTFaceTracker::StartTracking`	`const FT_SENSOR_DATA* data,`
	
	`const RECT* roi,`
	
	`const FT_VECTOR3D headPoints[2],`
	
	`IFTResult* result`
		Starts a new face tracking process with input sensor data, optional region of interest `roi`, optional `headPoints` hints, and writes out the `result`.

`IFTFaceTracker::ContinueTracking`	`const FT_SENSOR_DATA* data,`
	
	`const FT_VECTOR3D headPoints[2],`
	
	`IFTResult* result`
		Continues the face tracking process with input sensor data, optional `headPoints` hints, and writes out the `result`.

`IFTResult::GetFaceRect`	`RECT* rect`	Gets a face rectangle `rect` in the image frame resolution.

Please note that the face tracker may be affected by different factors. For example, if the light is too dark, or the player is wearing glasses, or he/she is too far away from the sensor, the tracking process may not be able to catch stable results.

Also note that the Kinect SDK's face tracking feature can support multiple face detection. We could just run the `StartTracking()` function with a `headPoints` parameter to indicate the head and the neck points, so that the face tracker can quickly find a face in a specific area. But, because the Kinect skeleton only supports tracking of not more than two people at any given point of time, the face tracker will hardly support more faces unless we specify the hint points by ourselves.

Constructing the face model

Now we will extend the previous example to support rendering the face mesh onto the color image. The face mesh itself may not be useful for your own applications, but it is in fact composited by a few parameterized points, that is, animation units (AU) and shape units (SU). The face tracking results can also be represented in terms of the weights of six AUs and 11 SUs, which will be explained in the forthcoming *Understanding the code* section.

Drawing the parametric face model

Now we are going to draw a 3D mesh of the face we have detected, which is formed by a list of vertices and normals, as well as the triangle indices. The Microsoft Kinect SDK already provides SDKs to obtain this data and this example will only draw them on the screen with OpenGL commands.

1. Common mesh data contains two parts: the vertices on the mesh, and the triangles composing the whole mesh, each of which includes three vertices. We define two variables at first to store the vertex data and the indices array representing all the triangles.

   ```
   std::vector<GLfloat> faceVertices(1);
   std::vector<GLuint> faceTriangles(1);
   ```

2. An `obtainFaceModelData()` function will be used to obtain all parameters related with the face mesh, including the positions, rotations, key vertices on the face for reconstructing the face model, and final mesh points and primitives for direct rendering.

   ```
   void obtainFaceModelData()
   {
       // See if we could get the face model for use
       IFTModel* model = NULL;
       HRESULT hr = tracker->GetFaceModel( &model );
       if ( FAILED(hr) ) return;

       // Obtain AUs and SUs of the face
       FLOAT* auList = NULL;
       UINT numAU = 0;
       if ( FAILED(faceResult->GetAUCoefficients(&auList,
   &numAU)) )
       {
   ```

```
        model->Release();
        return;
    }

    FLOAT* suList = NULL;
    UINT numSU = 0;
    BOOL haveConverged = FALSE;
    if ( FAILED(tracker->GetShapeUnits(NULL, &suList,
&numSU, &haveConverged)) )
    {
        model->Release();
        return;
    }

    // Obtain face model position, rotation and scale
    FLOAT scale, rotation[3], pos[3];
    if ( FAILED(hr = faceResult->Get3DPose(&scale,
rotation, pos)) )
    {
        model->Release();
        return;
    }

    // Declare variables to save face vertices and
triangles
    FT_TRIANGLE* triangles = NULL;
    UINT numTriangles = 0, numVertices = model-
>GetVertexCount();
    std::vector<FT_VECTOR2D> points2D( numVertices );

    // Obtain face model vertices and triangles
    POINT viewOffset = {0, 0};
    FT_CAMERA_CONFIG colorConfig = {640, 480,
NUI_CAMERA_COLOR_NOMINAL_FOCAL_LENGTH_IN_PIXELS};
    if ( SUCCEEDED(model->GetTriangles(&triangles,
&numTriangles)) &&
         SUCCEEDED(model->GetProjectedShape(&colorConfig,
1.0, viewOffset,
                    suList, numSU, auList, numAU, scale,
rotation, pos, &(points2D[0]), numVertices)) )
    {
        // The vertices are in camera coordinates,
        // so we have to re-projct them to [0, 1]
```

```
        faceVertices.resize( 3 * numVertices );
        for ( unsigned int i=0; i<numVertices; ++i )
        {
            faceVertices[3*i+0] = points2D[i].x / 640.0f;
            faceVertices[3*i+1] = 1.0f - (points2D[i].y /
480.0f);
            faceVertices[3*i+2] = 0.1f;
        }

        // Directly push each triangle indices to our
        // global index array
        faceTriangles.resize( 3 * numTriangles );
        for ( unsigned int n=0; n<numTriangles; ++n )
        {
            faceTriangles[3*n+0] = triangles[n].i;
            faceTriangles[3*n+1] = triangles[n].j;
            faceTriangles[3*n+2] = triangles[n].k;
        }
    }

    // Release the face model when everything is done
    model->Release();
}
```

3. In the `render()` function, we will try to obtain the face mesh data and render it using our convenient `drawIndexedMesh()` function. We change the OpenGL polygon mode to `GL_LINE` so that the model is drawn in the wireframe mode.

```
if ( isFaceTracked )
    obtainFaceModelData();

glDisable( GL_TEXTURE_2D );
glPolygonMode( GL_FRONT_AND_BACK, GL_LINE );

VertexData faceMeshData = { &(faceVertices[0]), NULL, NULL,
NULL };
drawIndexedMesh( WITH_POSITION, faceVertices.size()/3,
faceMeshData, GL_TRIANGLES, faceTriangles.size(),
&(faceTriangles[0]) );

glPolygonMode( GL_FRONT_AND_BACK, GL_FILL );
```

4. The rendering result, when we run the program, is shown in the following screenshot. We can see a very nice parametric face model shown when the Kinect SDK detects one face in the camera.

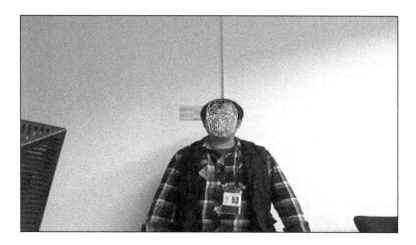

The facial mesh fitting with the face in real time

Understanding the code

The functions and methods used in this example are listed in the following table:

Function/method name	Parameters	Description
IFTFaceTracker::GetFaceModel	IFTModel** model	Returns an IFTModel to the face model.
IFTFaceTracker::GetShapeUnits	FLOAT** ppSUCoefs, UINT* pSUCount, BOOL* pHaveConverged	Returns shape units (SUs) and numbers in use: ppSUCoefs is the shape unit coefficients, pSUCount is the number, and pHaveConverged can be used to determine if SUs converge to real values.
IFTResult::GetAUCoefficients	FLOAT** ppCoefficients, UINT *pAUCount	Returns animation units (AUs) and numbers in use: ppCoefficients is the shape unit coefficients, pAUCount is the number.
IFTResult::Get3DPose	FLOAT* scale, FLOAT rotationXYZ[3], FLOAT translationXYZ[3]	Gets the 3D pose of the face model including translations, rotations, and scale.
IFTModel::GetVertexCount		Gets the number of vertices of the model.
IFTModel::GetTriangles	FT_TRIANGLE** ppTriangles, UINT* pTriangleCount	Gets mesh triangles of the 3D face model, and stores them in ppTriangles, with the triangle number stored in pTriangleCount.

Function/method name	Parameters	Description
IFTModel::GetProjectedShape	const FT_CAMERA_ CONFIG* cam, FLOAT zoomFactor, POINT viewOffset, const FLOAT *pSUCoefs, UINT suCount, const FLOAT *pAUCoefs, UINT auCount, FLOAT scale, const FLOAT rotationXYZ[3], const FLOAT translationXYZ[3], FT_VECTOR2D *pVertices, UINT vertexCount	• A face model is created with shape units, animation units, scale, rotation, and translation. Here, it is projected to the image frame space with the configuration cam. • The viewOffset and zoomFactor can be set to the same color as in initializeFaceTracker(). The following few parameters will be used to pass SU and AU points, 3D pose information. The final output shows pVertices (2D points) and vertexCount (number of vertices).

Here, we have mentioned the words "shape unit" (SU) and "animation unit" (AU) multiple times. They are actually derived from a classic parameterized face model named CANDIDE-3 (which is also the prototype of the Microsoft face tracker model). The original website for this is:

http://www.icg.isy.liu.se/candide/

In short, shape units (SUs) define the deformation of a standard face to a current player face, and animation units (AUs) define the delta values from the neutral shape to a morphed one. These two coefficients are very useful for emotion detection and recognition. For example, when the value of AU 4 is 1, the lip corners go up and thus make a very happy face; but when the value of AU 4 is -1, the lip corners are down to the bottom, so it means a sad face. More information about AU and SU computations and usage can be found at the CANDIDE website or at the Microsoft Face Tracking programming guide website:

http://msdn.microsoft.com/en-us/library/jj130970.aspx

Summary

In this chapter, we introduced the skeleton mapping technique in Kinect, and discussed how to get the positions of all the 20 joints in a human skeleton. Some of the positions are important for our successive developments; for example, the hand positions will be used to determine if a fruit is cut or not. Some more joint data will be used in the next chapter for emulating a multi-touch environment.

The new face tracking feature of the Microsoft Kinect SDK is also shown here with two easy-to-understand examples. The face tracking API can be used to calculate the position of a human head, as well as the mesh data composited from AUs and SUs. These two examples have nothing to do with the Fruit Ninja game in processing, but may be very useful for other kinds of AR-based applications and games.

5
Designing a Touchable
User Interface

In this chapter, we will introduce how to use Kinect APIs to simulate multitouch inputs, which are very common in modern interactive applications, especially on mobile devices. As a replacement of traditional methods (keyboard and mouse), the user interface of a multitouch-based application should always be dragged, or held and swiped, to trigger some actions. We will introduce some basic concepts of such interactions and demonstrate how to emulate them with Kinect.

Multitouch systems

The word multitouch refers to the ability to distinguish between two or more fingers touching a touch-sensing surface, such as a touch screen or a touch pad. Typical multitouch devices include tablets, mobile phones, pads, and even powerwalls with images projected from their back.

A single touch is usually done by a finger or a pen (stylus). The touch sensor will detect its X/Y coordinates and generate a touch point for user-level applications to use. If the device simultaneously detects and resolves multiple touches, user applications can thus efficiently recognize and handle complex inputs.

Gestures also play an important role in multitouch systems. A gesture is considered as a standardized motion, which can be used distinctly to represent a certain purpose. For example, the "tap" gesture (hit the surface lightly and release) always means to select and start a program on mobile phones, and the "zoom" gesture (move two fingers towards or apart from each other), or sometimes called the "pinch", is used to scale the content we are viewing.

Locating the cursors

In the first example of this chapter, we will convert the two hand bones into cursors to simulate a multitouch-like behavior. While the hand positions are changing, the cursors will also move so that we can locate them on a certain object, such as a button or a menu item. The available range of the hand positions must be limited here, otherwise the result will be confusing if we drop the arms and don't want the cursor to move again.

Drawing cursors from two hands

The line-trailing example in the previous chapter is a good start for our new task, so we will work on this example to add cursor support based on user-skeleton data we have already obtained. The steps are given as follows:

1. Define arrays to store necessary hand positions, as well as the colors to display in the window. We also declare a smoothParams variable here, which will be introduced later.

```
NUI_TRANSFORM_SMOOTH_PARAMETERS smoothParams;
GLfloat cursors[6];  // Left hand: 0, 1, 2; Right: 3, 4, 5
GLfloat cursorColors[8];  // Left hand: 0-3; Right: 4-7
```

2. In the update() function, add a line calling NuiTransformSmooth() to smooth the skeleton before using it.

```
NUI_SKELETON_FRAME skeletonFrame = {0};
hr = context->NuiSkeletonGetNextFrame( 0, &skeletonFrame );
if ( SUCCEEDED(hr) )
{
    context->NuiTransformSmooth( &skeletonFrame,
&smoothParams );
    for ( int n=0; n<NUI_SKELETON_COUNT; ++n )
    {
        NUI_SKELETON_DATA& data =
skeletonFrame.SkeletonData[n];
        if ( data.eTrackingState==NUI_SKELETON_TRACKED )
        {
            updateSkeletonData( data );
            break;
        }
    }
}
```

3. Rewrite the `updateSkeletonData()` function from the previous examples. We will only consider two hand bones and treat them as two 2D cursors on a virtual multitouch surface.

```
POINT coordInDepth;
USHORT depth = 0;
GLfloat yMin = 0.0f, zMax = 0.0f;
for ( int i=0; i<NUI_SKELETON_POSITION_COUNT; ++i )
{
    NuiTransformSkeletonToDepthImage(
        data.SkeletonPositions[i],
        &coordInDepth.x, &coordInDepth.y,
        &depth, NUI_IMAGE_RESOLUTION_640x480 );

    if ( i==NUI_SKELETON_POSITION_SPINE )
    {
    // Use the spine position to decide the available
    // cursor range
        yMin = 1.0f - (GLfloat)coordInDepth.y / 480.0f;
        zMax = (GLfloat)NuiDepthPixelToDepth(depth) *
0.00025f;
    }
    else if ( i==NUI_SKELETON_POSITION_HAND_LEFT )
    {
    // Obtain left hand cursor
        cursors[0] = (GLfloat)coordInDepth.x / 640.0f;
        cursors[1] = 1.0f - (GLfloat)coordInDepth.y /
480.0f;
        cursors[2] = (GLfloat)NuiDepthPixelToDepth(depth) *
0.00025f;
    }
    else if ( i==NUI_SKELETON_POSITION_HAND_RIGHT )
    {
    // Obtain right hand cursor
        cursors[3] = (GLfloat)coordInDepth.x / 640.0f;
        cursors[4] = 1.0f - (GLfloat)coordInDepth.y /
480.0f;
        cursors[5] = (GLfloat)NuiDepthPixelToDepth(depth) *
0.00025f;
    }
}
```

```
// If cursors are in range, show solid colors
// Otherwise make the cursors translucent
if ( cursors[1]<yMin || zMax<cursors[2] )
    cursorColors[3] = 0.2f;
else
    cursorColors[3] = 1.0f;
if ( cursors[4]<yMin || zMax<cursors[5] )
    cursorColors[7] = 0.2f;
else
    cursorColors[7] = 1.0f;
```

4. Render the cursors in the `render()` function. It is enough to represent these two cursors with two points, and the `glPointSize()` function to specify the point size on screen.

```
glDisable( GL_TEXTURE_2D );
glPointSize( 50.0f );

VertexData cursorData = { &(cursors[0]), NULL,
&(cursorColors[0]), NULL };
drawSimpleMesh( WITH_POSITION|WITH_COLOR, 2, cursorData,
GL_POINTS );
```

5. Don't forget to initialize the smoothing variable and the colors in the main entry. The color is all set to white here.

```
smoothParams.fCorrection = 0.5f;
smoothParams.fJitterRadius = 1.0f;
smoothParams.fMaxDeviationRadius = 0.5f;
smoothParams.fPrediction = 0.4f;
smoothParams.fSmoothing = 0.2f;
for ( int i=0; i<8; ++i ) cursorColors[i] = 1.0f;
```

6. Run the program and we can see two circles (in fact, points) on the screen instead of line trails. When we put our hands behind our body, or down to the hip, the points turn translucent. And if we lift any of the arms in front of the chest, the circle will become opaque to indicate that the cursor is valid now.

7. The snapshot of this example is shown in the following image:

Rendering the cursors from the hands

Understanding the code

Kinect will always return all the bones' data for every frame to user applications, no matter whether this data is being currently tracked or inferred from the previous frame data, so we have to manually set a range within which user motions can be parsed to 2D cursors on a virtual surface. An arbitrary range may be inconvenient for real use. For example, a gesture may be triggered unexpectedly when the user already puts his/her hands down and wants to have a look at the current scene.

As a demo here, we will set a very simple limitation that uses the spine location as the datum point. When the hand location in the camera space is behind the spine (on the z axis), or lower than the spine (on the y axis), we think that it is invalid and make it translucent; otherwise, it is in use and can trigger interface events, if we have any.

A new NUI function is also introduced in this example:

```
HRESULT NuiTransformSmooth(
    NUI_SKELETON_FRAME* skeletonFrame,
    const NUI_TRANSFORM_SMOOTH_PARAMETERS* smoothingParams
);
```

It reads skeleton positions from `skeletonFrame` and reduces jitters of them, according to the parameters provided in `smoothingParams`. The skeleton data in tracking may always be inaccurate because of the capturing and computing precisions, or just clipped by some occludes. In such a case, a bone may jitter because the inferred values of two frames are obviously different. The `NuiTransformSmooth()` function can reduce such a problem here, with the cost that the result may have latency compared with the actual motions.

The parameters of `NuiTransformSmooth()` are defined in the following structure:

```
typedef struct _NUI_TRANSFORM_SMOOTH_PARAMETERS
{
    FLOAT fSmoothing;  // Smoothing parameter in [0, 1]
    FLOAT fCorrection;  // Correction parameter in [0, 1]
    FLOAT fPrediction;  // Number of frames to predict
    FLOAT fJitterRadius;  // Jitter-reduction radius, in meters
    FLOAT fMaxDeviationRadius;  // Max radius of filtered
positions
} NUI_TRANSFORM_SMOOTH_PARAMETERS;
```

Additional information

You may edit the smoothing parameters of this example and see if anything has changed. Remember that higher smoothing leads to higher latency, and better jitter-reduction means lower positioning accuracy. It is at your own risk to use or overuse this feature in your applications.

Common touching gestures

The next step of this chapter is to support some very basic gestures so our cursors can really work in interactive applications, rather than only provide the locations. Before that, we will first introduce common single and multitouch gestures and how they are implemented in this section. Although we are going to finish only two of them (holding and swiping), it is still necessary to have a general understanding here, for the purpose of developing a gesture-based user interface in the future.

Gesture name	Action	Equivalent mouse action
Tap	Press on the surface lightly.	Click a button.
Double tap	Tap twice on the surface.	Double click on a program icon and start it.
Hold	Press on the surface and wait for a while.	Simulates right-clicking on touch screens.

Gesture name	Action	Equivalent mouse action
Swipe	Drag on the surface and release quickly.	Pans the scroll bars to view parts of the content.
Drag	Drag slowly on the surface.	Drags an item and drop it somewhere.
Two-finger tap	Click on the surface with two fingers at the same time.	None.
Zoom/Pinch	Move two fingers on the surface, towards or apart from each other.	Simulates the mouse wheel on touch screens.
Rotate	Make one finger the pivot, and move another around.	None.
	Another implementation is to move the two fingers in opposing directions.	

There are more resources about multitouch gesture implementations in depth. For example, the Microsoft Touch Gesture website:

```
http://msdn.microsoft.com/en-us/library/windows/desktop/
dd940543(v=vs.85).aspx
```

You can also see the wiki page, which explains the history and implementation details:

```
http://en.wikipedia.org/wiki/Multi_touch
```

Recognizing holding and swiping gestures

We are going to implement two of the gestures from the previous table in this section: holding and swiping. They are actually not "multi" gestures because they can be finished with only one finger on the surface, or one hand cursor from Kinect, but both are very useful for developing Kinect-based applications. The holding gesture can be used to trigger a button on the screen, and the swiping gesture can be used to select the menu item, by scrolling the item list and finding the required one.

Drawing cursors using two hands

Let's start now.

1. Declare variables for gesture recognizing. We can increase the related counters when prerequisites of a specific gesture are fitted. And when the counter reaches a certain value, we will mark the gesture as "recognized".

    ```
    GLfloat cursors[6];   // Left hand: 0, 1, 2; Right: 3, 4, 5
    GLfloat lastCursors[6];
    GLfloat cursorColors[8];   // Left hand: 0-3; Right: 4-7
    unsigned int holdGestureCount[2] = {0};
    unsigned int swipeGestureCount[2] = {0};
    ```

2. At the end of the `updateSkeletonData()` function, we declare and call a new function named `guessGesture()`. It will check possible gestures of both cursors. The locations will then be recorded to the variable `lastCursors` for the next frame use.

    ```
    guessGesture( 0, (yMin<cursors[1] && cursors[2]<zMax) );
    guessGesture( 1, (yMin<cursors[4] && cursors[5]<zMax) );
    for ( int i=0; i<6; ++i ) lastCursors[i] = cursors[i];
    ```

3. The `guessGesture()` function has two parameters: the cursor index (0 is the left hand, and 1 is the right hand), and a Boolean value to tell if the cursor is in the available range.

    ```
    void guessGesture( unsigned int index, bool inRange )
    {
    ...
    }
    ```

4. In the function body, we will determine if the current state fits the conditions of either holding or swiping a gesture. Because `lastCursors` records the cursor locations of the previous frame, we can obtain the velocities of both cursors between two frames and use them for instantaneous judgment.

    ```
    if ( !inRange )
    {
        // If the hand is not in range, reset all counters and
        // the cursor color (turn to translucence)
        holdGestureCount[index] = 0;
        swipeGestureCount[index] = 0;
        cursorColors[3 + index*4] = 0.2f;
    }
    else
    {
    ```

```
// Compute the distance of this and last cursor, which
// is actually the instantaneous velocity of the cursor
float distance = sqrt(
    powf(cursors[index*3]-lastCursors[index*3], 2.0f) +
    powf(cursors[1+index*3]-lastCursors[1+index*3],
2.0f));
    if ( distance<0.02 )
    {
        // If the cursor is nearly unmoved, increase
holding
        holdGestureCount[index]++;
        swipeGestureCount[index] = 0;
    }
    else if ( distance>0.05 )
    {
        // If the cursor changes fast, increase swiping
        holdGestureCount[index] = 0;
        swipeGestureCount[index]++;
    }
    else
    {
        // Otherwise, reset the counters
        holdGestureCount[index] = 0;
        swipeGestureCount[index] = 0;
    }
    cursorColors[3 + index*4] = 1.0f;
}
```

5. We will print the gesture names on screen in the render() function. If the counters for holding gesture are increased to a large enough value (in this case, to 30), it means we are holding the cursor for a long time, so "hold" is triggered. And if the swiping counters are set, it means we have already swiped the hands and the "swipe" gesture is triggered.

```
std::string text = "Gestures (L/R): ";
for ( int i=0; i<2; ++i )
{
    if ( holdGestureCount[i]>30 ) text += "Hold;";
    else if ( swipeGestureCount[i]>1 ) text += "Swipe;";
    else text += "None;";
}
glRasterPos2f( 0.01f, 0.01f );
glColor4f( 1.0f, 1.0f, 1.0f, 1.0f );
glutBitmapString( GLUT_BITMAP_TIMES_ROMAN_24, (const
unsigned char*)text.c_str() );
```

6. Now run the application and try to achieve these gestures by using your hands. The gesture names will be displayed at the bottom-left corner of the screen. Now lift one of your arms and stop for some time, and flick the other hand quickly in front of you. Pay attention to the text displayed on the screen and see if the recognized gestures are correct and stable.

Gestures (L/R): Hold;Swipe;

Display hand gestures (at the bottom-left)

7. Please note that the swiping gesture may not be easily noticed from the text. That's because it will happen in a very short time. The next example of this chapter will demonstrate the use of the swiping gesture more clearly.

Understanding the code

Determining gestures always requires some mathematics, as well as some user-engineering knowledge. An instantaneous state may not be used to decide if a gesture happens or not. For example, the holding gesture requires the user to keep his/her cursor at a specified place for a while, and double tapping means we must detect the "tap" gesture at least twice in a short period of time. Thus, we have to keep a historical list of the cursors, which is similar to the implementation of a linetrail effect in the previous chapter.

Here are some hints for implementing different gestures, including the two gestures we have already done:

- **Tap**: Not good for Kinect-based use, as you can hardly decide the time of "pushing" and "releasing" motions.

- **Double tap**: Again, not good for Kinect-based use, as you can hardly decide the time of "pushing" and "releasing" motions.

- **Hold**: Checks the distance between the current cursor and the previous cursor for every frame to see if they are near enough. The holding gesture is triggered if the cursor is still for a significant amount of time.

- **Swipe**: Checks the distance between the current cursor and the previous cursor. If the distance is large enough, the user must exert himself to fling the arms and thus make a "swiping" gesture. Note that you must exclude the jitters.

- **Drag**: Checks the distance between the current cursor and the previous cursor for every frame to determine if the cursor is moving all the time, while neither exceeding the holding and swiping threshold.

- **Zoom**: Checks the distance of both the cursors. If the average velocities are opposite and the historical cursors of both are in a line, it produces a "zooming" gesture.

- **Rotate**: Checks the distance of both the cursors. Make sure that the historical cursors of both are not in a line. If one is still and the other is moving a lot, or the average velocities are opposite, it can be considered as a "rotating" gesture.

Maybe you will have some other ideas and solutions, so don't hesitate to replace any in the previous list with your own, and see if it can make your customers feel better.

Additional information

Try to implement some more gestures on your own, especially the dragging and zooming gestures. They are very useful in your future projects for easy user interactions.

Sending cursors to external applications

In the last example of this chapter, we are going to make some use of the cursors we obtained from the skeleton data. As there are many other applications developed with only the mouse and keyboard as main input devices, it is sometimes meaningful to synthesize keyboard inputs, mouse motions, and button clicks, and send them to these applications for the purpose of providing more interaction methods.

A cool example, which we will be implementing here, is to convert the cursor data from Kinect to Windows mouse information so that we can use motion-sense techniques to control common operation systems. Other useful ideas include converting the cursors to the TUIO protocol (a unified framework for tangible multitouch surfaces) and use them for remote controls, or communicating with some famous multimedia software such as VVVV and Max/MSP.

Emulating Windows mouse with cursors

This recipe will be slightly different from the previous ones. It won't render anything in the OpenGL window, but will send mouse events to Windows, to control the real mouse cursor. This requires us to make use of the Windows function SendInput().

1. We don't have to make it fullscreen. Just comment the following line:

   ```
   //glutFullScreen();
   ```

2. At the end of updateSkeletonData(), after guessing the possible gestures, we convert the left hand's location and gesture into mouse press and wheel events, with the help of the Windows system API SendInput().

   ```
   if ( cursors[2]<zMax )
   {
       // Set values of the INPUT structure
       INPUT input = {};
       input.type = INPUT_MOUSE;
       input.mi.dx = (LONG)(65535 * cursors[0]);
       input.mi.dy = (LONG)(65535 * (1.0 - cursors[1]));
       input.mi.dwExtraInfo = GetMessageExtraInfo();

       if ( holdGestureCount[0]>30 )
       {
           // Send mouse push and release events when the
           // holding gesture is detected
           if ( !isMouseDown )
           {
   ```

```
            input.mi.dwFlags = MOUSEEVENTF_ABSOLUTE |
MOUSEEVENTF_LEFTDOWN;
        }
        else
        {
            input.mi.dwFlags = MOUSEEVENTF_ABSOLUTE |
MOUSEEVENTF_LEFTUP;
        }
        isMouseDown = !isMouseDown;

        // Reset the counter so we won't receive the same
        // gesture continuously
        holdGestureCount[0] = 0;
    }
    else if ( swipeGestureCount[0]>1 )
    {
        // If we encounter the swiping gesture, use it
        // to emulate the mouse wheel to scroll pages
        input.mi.dwFlags = MOUSEEVENTF_WHEEL;
        input.mi.mouseData = WHEEL_DELTA;
    }
    else
    {
        // For all other cases, simply move the mouse
        input.mi.dwFlags = MOUSEEVENTF_ABSOLUTE |
MOUSEEVENTF_MOVE;
    }
    SendInput( 1, &input, sizeof(INPUT) );
}
```

3. Start the program and stand in front of the Kinect device. Now lift the left hand and we can see the Windows mouse cursor also move along with us. Hold and stay on an icon on the desktop. This will be recognized as "selecting and dragging", and a quick swipe in any direction will result in scrolling the content we are currently viewing.

Understanding the code

There is nothing special in this example. We just used the Windows native function SendInput() to send cursors and gestures inferred from Kinect data to Windows mouse events. Windows uses the top-left corner as the original point, so we have to alter the coordinates before sending them. The gestures are parsed as left clicks and wheel events here, but of course you can consider them as different operations.

It is more valuable to change your Kinect cursors to TUIO, which can then be sent to local or remote clients that listen to the current TUIO server. There are several pieces of multimedia software that regard TUIO as a regular input, so you can easily connect them with Kinect devices.

The TUIO website is:

```
http://www.tuio.org/
```

TUIO API for C/C++ and other languages can be found at:

```
http://www.tuio.org/?software
```

Simply call the corresponding TUIO commands when you receive a new cursor position or detect a new gesture, and any clients that listen to your PC using the same TUIO protocol will receive it and can parse it for their own uses.

Summary

We introduced how to emulate mouse cursors and some single and multitouch behaviors from skeleton inputs. This is useful for developing the user interface of a Kinect-based application, which is hands-free and can hardly benefit from common interaction methods.

Using gestures such as holding and swiping, it is now possible to add buttons, menu items, and other triggerable elements in our application and select them with definite and exclusive motions. And this will also help develop the graphics interface of the Fruit Ninja game in the next chapter.

Another important highlight of this chapter is the integration of Kinect inputs and applications without Kinect supports. We use the Windows system API to emulate mouse inputs in this chapter. But it is also suggested to make use of the TUIO protocol to be compatible with more multitouch applications in the future.

6
Implementing the Scene and Gameplay

It may be a long time before we really get ready to create our Fruit Ninja game, which is Kinect-based, with some augmented reality features. We have already obtained the knowledge of building complete Kinect applications with image streaming, skeleton tracking, and simple multi-touch gesture support. Now, in this chapter we will integrate all that we learnt together to quickly finish the basic elements of our game, and add simple game logic that includes computing the player scores and dynamically changing difficulty levels.

We will not include all the components of a complete commercial game because of the page and resource limitation. Good art design (graphics, sound, and so on) is always important for a game to grasp the consumer's attention at first sight. Other features, including network and multiplayer supports, will highly increase the replay-ability of the game. But these are not necessary for this book to introduce and implement, as they have no relationship with Kinect's programming techniques.

Integrating the current code

Until now, we have successfully created examples about displaying a translucent depth image on a background image (*Chapter 3, Rendering the Player*), drawing linetrails of the two hands (*Chapter 4, Skeletal Motion and Face Tracking*), and determining a simple holding gesture (*Chapter 5, Designing a Touchable User Interface*) of UI interaction. These can just be used in the Fruit Ninja game to display the player and emulate his/her blade paths. The most basic UI components of our game are the restart and exit buttons. We will use the holding gesture to check if the player clicks on either of the two buttons.

We use an image with two buttons (restart and exit) drawn at the top-left and top-right corners of the background. We will use the holding gesture to trigger the buttons. The image is shown as follows:

The background image

Integrating existing elements in a scene

Now let's start.

1. The following global variables are copied and altered from the examples in *Chapter 3*, *Rendering the Player*, *Chapter 4*, *Skeletal Motion and Face Tracking*, and *Chapter 5*, *Designing a Touchable User Interface*. They are mainly used for three different purposes: displaying the player image, displaying hand point and trails, and checking possible gestures.

```
INuiSensor* context = NULL;
HANDLE colorStreamHandle = NULL;
HANDLE depthStreamHandle = NULL;
TextureObject* playerDepthTexture = NULL;

struct Vertex { GLfloat x, y, z; };
std::vector<Vertex> leftHandTrails;
std::vector<Vertex> rightHandTrails;
GLfloat trailColors[20][4];
```

```
NUI_TRANSFORM_SMOOTH_PARAMETERS smoothParams;
unsigned int holdGestureCount[2] = {0};
const unsigned int backgroundTexID = 1;
```

2. We will modify the code in guessGesture() to compute the distance between the first and last points in the historical list. The resulting gesture is a holding gesture if the distance is small enough in a span.

```
float distance = 0.0f, currentX = 0.0f, currentY = 0.0f;
if ( index==0 )   // left hand
{
    currentX = leftHandTrails.back().x;
    currentY = leftHandTrails.back().y;
    distance = sqrt(
        powf(currentX-leftHandTrails.front().x, 2.0f)
      + powf(currentY-leftHandTrails.front().y, 2.0f));
}
else  // right hand
{
    currentX = rightHandTrails.back().x;
    currentY = rightHandTrails.back().y;
    distance = sqrt(
        powf(currentX-rightHandTrails.front().x, 2.0f)
      + powf(currentY-rightHandTrails.front().y, 2.0f));
}

.// Increase the holding count if distance is small enough
if ( distance<0.02 )
    holdGestureCount[index]++;
else
    holdGestureCount[index] = 0;

// The player is holding something, check if he pressed
// one of the buttons (at top-left or top-right)
if ( holdGestureCount[index]>30 )
{
    if ( currentY>0.9f && currentX<0.1f )   // Restart
    { /* do nothing at present */ }
    else if ( currentY>0.9f && currentX>0.9f )   // Exit
        glutLeaveMainLoop();
}
```

3. Finally, in the `updateSkeletonData()` function, check the gesture states as follows:

```
guessGesture( 0, (yMin<leftHandTrails.back().y &&
leftHandTrails.back().z<zMax) );
guessGesture( 1, (yMin<rightHandTrails.back().y &&
rightHandTrails.back().z<zMax) );
```

4. And in the main entry, we will successively initialize the background texture, the trail attributes, the Kinect device and the player image data, and the smoothing parameters.

```
if ( TextureManager::Inst()->LoadTexture("FruitNinja1.jpg",
backgroundTexID, GL_BGR_EXT) )
{
    glTexParameteri( GL_TEXTURE_2D, GL_TEXTURE_MIN_FILTER,
GL_LINEAR );
    glTexParameteri( GL_TEXTURE_2D, GL_TEXTURE_MAG_FILTER,
GL_LINEAR );
}

for ( int i=0; i<20; ++i )
{
    trailColors[i][0] = 1.0f;
    trailColors[i][1] = 1.0f;
    trailColors[i][2] = 1.0f;
    trailColors[i][3] = (float)(i + 1) / 20.0f;
}

if ( !initializeKinect() ) return 1;
playerDepthTexture = createTexture(640, 480,
GL_LUMINANCE_ALPHA, 2);

smoothParams.fCorrection = 0.5f;
smoothParams.fJitterRadius = 1.0f;
smoothParams.fMaxDeviationRadius = 0.5f;
smoothParams.fPrediction = 0.4f;
smoothParams.fSmoothing = 0.2f;
```

5. Start the program, and we will see the depth image, the handtrails, and the background image as expected.

The Fruit Ninja game interface

6. Isn't it cool? Now you can also hold your right hand on the close button at the top-right corner to quit the program.

Understanding the code

Originally, the Fruit Ninja game was very easy to understand and it didn't need a Kinect device at all. It only required the player to drag his fingers on the pad screens and slice as many fruits as possible to earn higher scores. There were also some occasional bombs hidden in the fruits thrown from the bottom of the screen, so the player must be careful to avoid them; otherwise he will lose his life and the game is over. The faster the fruits and bombs are thrown up, the more difficult the game becomes.

Using a Kinect device will also increase the difficulty of the game. That's because the player must swing his arms to make the blade move. The accuracy of hitting a fruit will also be lower than that of using a finger or a mouse. In this example, we use the depth image to indicate the player position and shape so that he/she won't be blinded when facing the screen. The skeleton tracking feature is used only for obtaining the hand positions and is a must-have element of this game as it determines if the player reaches any fruit object and earns new scores, or holds on an exit button to quit to the desktop.

Cutting the fruits

Now we really want to "cut" something instead of just seeing the trailing effect. To simplify the process, we should have the following prerequisite conditions:

- The fruit and bomb objects are represented by a 2D RGBA image with a transparent background. The positions and states (whether sliced or not) are updated in every frame.

- The fruit is thrown from the bottom of the screen with a random initial velocity. It will fall down because of gravity.

- When the hand points reach the object image (in fact, a rectangle), it is divided into four parts as if sliced by a mystical ninja.

The code doesn't need to be related with Kinect, but we will have to pass the hand points and velocities to determine if a fruit object is cut. All these operations can be done in the `update()` function.

It is also important to find or create some fruit and bomb images (using cartoon styles may be a good idea). In this simple example, we choose the following ones:

Fruits/bomb images with a transparent background

Adding and handling fruit objects

Now let's start with the process of adding and handling fruit objects.

1. We use an independent class to declare all the properties and methods
    that a fruit object should have.

    ```cpp
    class FruitObject
    {
    public:
        // The constructor
        FruitObject( unsigned int id, bool b, GLfloat s,
                    GLfloat tx=0.0f, GLfloat ty=0.0f,
                    GLfloat tw=1.0f, GLfloat th=1.0f );

        // Update position and velocity of the fruit
        void update();

        // Render the fruit object
        void render();

        // Size of the fruit image
        GLfloat size;

        // Offset of the texture coordinate for slicing use
        GLfloat texOffset[4];

        // Position of the fruit image
        Vertex position;

        // Velocity of the fruit image
        Vertex velocity;

        // Texture ID of the fruit image
        unsigned int objectID;

        // Flag to determine if the image can be sliced
        bool canSlice;
    };
    ```

2. We will maintain a list of the `FruitObject` variables, and define four
    different texture IDs (watermelon, apple, mango, and bomb) for use.

    ```cpp
    std::vector<FruitObject> _fruitObjects;
    const unsigned int objectTexIDs[4] = {2, 3, 4, 5};
    ```

3. Use a simple random value generating function to create a random value between `min` and `max`.

```
float randomValue( float min, float max )
{ return (min + (float)rand()/(RAND_MAX+1.0f)*(max - min)); }
```

4. In the `update()` function, we will have to perform two major tasks: first we have to traverse and update all existing fruit objects, remove objects out of the screen range, and slice objects into four smaller parts, and second, generate new fruits and shoot them from the bottom randomly.

```
// Store new objects temporarily
std::vector<FruitObject> newObjects;

// Traverse all existing fruits
for ( std::vector<FruitObject>::iterator
itr=_fruitObjects.begin(); itr!=_fruitObjects.end(); )
{
    FruitObject& fruit = (*itr);
    bool isSliced = false;
    if ( fruit.canSlice )
    {
        // Check distance between the fruit origin and
        // hand points. If near enough, mark as sliced
        float distance = sqrt(powf(fruit.position.x-
leftHandTrails.back().x, 2.0f) + powf(fruit.position.y-
leftHandTrails.back().y, 2.0f));
        if ( distance<fruit.size ) isSliced = true;

        distance = sqrt(powf(fruit.position.x-
rightHandTrails.back().x, 2.0f) + powf(fruit.position.y-
rightHandTrails.back().y, 2.0f));
        if ( distance<fruit.size ) isSliced = true;
    }

    if ( isSliced )
    {
        // Slice into 4 parts and remove the old one
        // Will explain slicing process in the next step
        ...
        itr = _fruitObjects.erase( itr );
    }
    else if ( fruit.position.y<0.0f )
    {
```

```
                // If object is under the screen bottom, remove it
                itr = _fruitObjects.erase( itr );
            }
            else
            {
                // For all alive fruit objects, update them
                fruit.update();
                ++itr;
            }
        }
    }
```

5. Slice an existing object into four parts. A very easy-to-understand way to do this is to divide the original image into four equal parts with the same size, using the texture coordinate offsets to determine its display. Then we add an offset to the current velocity to separate these four parts from each other.

```
float deltaX = fabs(fruit.velocity.x * 0.2f);
float deltaY = fabs(fruit.velocity.y * 0.2f);

// Note, smaller parts can't be sliced again
FruitObject chop1( fruit.objectID, false, 0.05f, 0.0f,
0.0f, 0.5f, 0.5f );
chop1.position.x = fruit.position.x;
chop1.position.y = fruit.position.y;
chop1.velocity.x = fruit.velocity.x - deltaX;
chop1.velocity.y = fruit.velocity.y - deltaY;
newObjects.push_back( chop1 );

FruitObject chop2( fruit.objectID, false, 0.05f, 0.5f,
0.0f, 0.5f, 0.5f );
chop2.position.x = fruit.position.x + fruit.size*0.5f;
chop2.position.y = fruit.position.y;
chop2.velocity.x = fruit.velocity.x + deltaX;
chop2.velocity.y = fruit.velocity.y - deltaY;
newObjects.push_back( chop2 );

FruitObject chop3( fruit.objectID, false, 0.05f, 0.5f,
0.5f, 0.5f, 0.5f );
chop3.position.x = fruit.position.x + fruit.size*0.5f;
chop3.position.y = fruit.position.y + fruit.size*0.5f;
chop3.velocity.x = fruit.velocity.x + deltaX;
chop3.velocity.y = fruit.velocity.y + deltaY;
newObjects.push_back( chop3 );
```

```
FruitObject chop4( fruit.objectID, false, 0.05f, 0.0f,
0.5f, 0.5f, 0.5f );
chop4.position.x = fruit.position.x;
chop4.position.y = fruit.position.y + fruit.size*0.5f;
chop4.velocity.x = fruit.velocity.x - deltaX;
chop4.velocity.y = fruit.velocity.y + deltaY;
newObjects.push_back( chop4 );
```

6. The next step is to randomly create new fruits with random initial positions and velocities. This is also decided by a random value (when it is less than 0.01).

```
bool createNew = randomValue(0.0f, 1.0f) < 0.01f;
if ( createNew )
{
    FruitObject obj( (int)randomValue(0.0f, 3.9f), true,
0.1f );
    obj.position.x = randomValue(0.1f, 0.9f);
obj.position.y = 0.0f;
    obj.velocity.x = randomValue(0.006f, 0.012f);
    obj.velocity.y = randomValue(0.03f, 0.04f);
    if ( obj.position.x>0.5f ) obj.velocity.x = -
obj.velocity.x;
    newObjects.push_back( obj );
}

// Insert new objects (sliced and newly created) into list
_fruitObjects.insert( _fruitObjects.end(),
newObjects.begin(), newObjects.end() );
```

7. In the `render()` function, we traverse the objects again and render them one by one.

```
for ( unsigned int i=0; i<_fruitObjects.size(); ++i )
    _fruitObjects[i].render();
```

8. Now, we will implement each method of the `FruitObject` class. In the constructor, we will receive the texture ID, the slicing flag, the size, and the texture coordinate offsets as inputs. Set them to member variables for use.

```
FruitObject::FruitObject( unsigned int id, bool b, GLfloat
s,
            GLfloat tx, GLfloat ty, GLfloat tw, GLfloat th
)
{
    canSlice = b; size = s;
    texOffset[0] = tx; texOffset[1] = ty;
    texOffset[2] = tx + tw; texOffset[3] = ty + th;
```

```
    position.x = 0.0f; position.y = 0.0f; position.z =
0.0f;
    velocity.x = 0.0f; velocity.y = 0.0f; velocity.z =
0.0f;
    objectID = id;
}
```

9. In the `FruitObject::update()` method, we add the current velocity to the position, and alter the velocity with an imaginary gravity pull.

```
void FruitObject::update()
{
    position.x += velocity.x;
    position.y += velocity.y;

    const GLfloat gravity = -0.001f;
    velocity.y += gravity;
}
```

10. In the `FruitObject::render()` method, we draw the fruit object's corresponding image onto a quad. Because this method is called in `render()` after enabling blending, the alpha channel of the image will be used, and background pixels with zero alpha values will be automatically culled.

```
void render()
{
    GLfloat vertices[][3] = {
        { 0.0f, 0.0f, 0.0f }, { size, 0.0f, 0.0f },
        { size, size, 0.0f }, { 0.0f, size, 0.0f }
    };
    GLfloat texcoords[][2] = {
        {texOffset[0], texOffset[1]}, {texOffset[2],
texOffset[1]},
        {texOffset[2], texOffset[3]}, {texOffset[0],
texOffset[3]}
    };
    VertexData meshData = { &(vertices[0][0]), NULL, NULL,
&(texcoords[0][0]) };

    glPushMatrix();
    glTranslatef( position.x, position.y, position.z );
    TextureManager::Inst()->BindTexture(
objectTexIDs[objectID] );
    drawSimpleMesh( WITH_POSITION|WITH_TEXCOORD, 4,
meshData, GL_QUADS );
    glPopMatrix();
}
```

11. Don't forget to load the four fruit images (starting with the filename from `FruitNinja2.png` to the filename `FruitNinja5.png`) in the main entry.

```
for ( int i=0; i<4; ++i )
{
    std::stringstream ss;
    ss << "FruitNinja" << i+2 << ".png";
    if ( TextureManager::Inst()-
>LoadTexture(ss.str().c_str(), objectTexIDs[i],
GL_BGRA_EXT, 4) )
    {
        glTexParameteri( GL_TEXTURE_2D,
GL_TEXTURE_MIN_FILTER, GL_LINEAR );
        glTexParameteri( GL_TEXTURE_2D,
GL_TEXTURE_MAG_FILTER, GL_LINEAR );
    }
}
```

12. Start the program. You will see fruits randomly jumping out from the bottom of the screen. Now swipe your arm to cut them!

Cut the fruits with Kinect-driven trails

Understanding the code

When the fruit image is near enough to any hand point, we treat it as sliced and will cut it into four parts. Although fragmentation is in fact a more complex process and requires a lot of physics computations, we will not make such mistakes in this book, and will divide the original image into four equal parts as shown in the following figure.

The new length of any part is just half of the original one. And the texture coordinate should also be translated and scaled. For example, the fragment at the bottom-left has new texture coordinates from (0, 0) to (0.5, 0.5), and the one at the top-right is from (0.5, 0.5) to (1, 1). So we must alter the input arguments of `FruitObject` while creating these sliced parts as shown in the following code:

```
// The new object has the same texture ID as the original one
// The size is just the half, and it can't be sliced again
// The new texture coordinates is from (0, 0) to (0.5, 0.5)
FruitObject chop1( fruit.objectID, false, 0.05f, 0.0f, 0.0f, 0.5f,
0.5f );
```

And the velocity of the fragments must be changed slightly and gradually to make the four parts move apart from each other. The following figure shows how an offset value is added to each new velocity to make it act like an explosion:

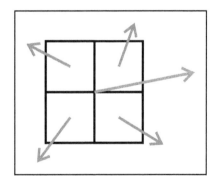

The original velocity (light blue) and the velocity offsets (green) adding to the four sliced parts

And it's simply done using the following code segments:

```
float deltaX = fabs(fruit.velocity.x * 0.2f);
float deltaY = fabs(fruit.velocity.y * 0.2f);
...
chop.velocity.x = fruit.velocity.x +/- deltaX;
chop.velocity.y = fruit.velocity.y +/- deltaY;
```

Playing the game

The last example in this chapter is to design a simple gameplay. We now have the interaction driven by Kinect, and fruit objects that intersect with handtrails, but we don't have any rules yet. The gameplay will specify how to win or lose the game and how to set challenges for the player to overcome. A good gameplay is the base of the playability of a game.

We have four rules for the Fruit Ninja game here to follow:

- Our goal is to cut as many fruits as possible to earn higher scores.
- The higher the points we earn, the faster and more in number the fruits appear.
- The player has a life value (initially 100). The game finishes when it becomes zero.
- If a fruit is not sliced, the life will be decreased by 5 points. If a bomb is sliced, the life will be decreased by 20 points.

And these rules will be implemented in the following section.

Adding simple game logic

Let's start now.

1. We will declare the `score` and `life` variables, which will act as the very basic elements of the gameplay. And we add a `gameOverTexID` variable for displaying a different image when the game is over (when the life is down to 0).

   ```
   const unsigned int gameOverTexID = 6;
   int score = 0, life = 100;
   ```

2. In `guessGesture()`, we will reset the `score` and `life` variables if the player holds his hand on the restart button (top-left), and quit the program when holding on the close button (top-right).

   ```
   if ( holdGestureCount[index]>30 )
   {
       if ( currentY>0.9f && currentX<0.1f )  // Restart
       { score = 0; life = 100; }
       else if ( currentY>0.9f && currentX>0.9f )  // Exit
           glutLeaveMainLoop();
   }
   ```

3. In the `update()` function, we have to check the `objectID` object of every fruit object intersected with the hands. If it is a bomb (ID = 3), decrease the life; otherwise increase the score.

```
if ( isSliced )
{
    . . .
    if ( fruit.objectID<3 ) score += 10;
    else life -= 20;
}
else if ( fruit.position.y<0.0f )
{
    // If the fruit is sliceable but not sliced
    // we also decrease the life as a punishment
    if ( fruit.canSlice ) life -= 5;
    itr = _fruitObjects.erase( itr );
}
else
{
    fruit.update();
    ++itr;
}
```

4. To make the game more difficult, when the player earns a high enough score, we can increase the chance of generating new fruits in every frame gradually. This is done by simply altering the following code:

```
bool createNew = randomValue(0.0f, 1.0f) < 0.01f * (1.0f +
(float)score / 100.0f);
if ( createNew )
{
    . .
}
```

5. In the `render()` function, we check if the life equals to 0 and display the "game over" texture instead of the normal one.

```
if ( life<=0 )
    TextureManager::Inst()->BindTexture( gameOverTexID );
else
    TextureManager::Inst()->BindTexture( backgroundTexID );
drawSimpleMesh( WITH_POSITION|WITH_TEXCOORD, 4, meshData,
GL_QUADS );
```

6. We can also write the current score and life values on the screen in the `render()` function.

```
std::stringstream ss;
ss << "Score: " << score << "  Life: " << (life<0? 0:
life);

glRasterPos2f( 0.01f, 0.01f );
glColor4f( 1.0f, 1.0f, 1.0f, 1.0f );
glutBitmapString( GLUT_BITMAP_TIMES_ROMAN_24, (const
unsigned char*)ss.str().c_str() );
```

7. Last but not least, read the "game over" image from the disk. You may design this image by adding some texts and symbols on the normal background.

```
if ( TextureManager::Inst()->LoadTexture("FruitNinja6.jpg",
gameOverTexID, GL_BGR_EXT) )
{
    glTexParameteri( GL_TEXTURE_2D, GL_TEXTURE_MIN_FILTER,
GL_LINEAR );
    glTexParameteri( GL_TEXTURE_2D, GL_TEXTURE_MAG_FILTER,
GL_LINEAR );
}
```

8. Now, compile and run our game. You will see the score and life values changing when you struggle to cut the fruits thrown suddenly.

Playing the Kinect-based Fruit Ninja game

9. When the game is over, you will see a different image on the screen, and you can hold on to the restart button at the top-left corner to restart the game, or exit by holding on to the exit button at the top-right corner.

The game is over if the value of the player's life reaches zero

Understanding the code

A perfect gameplay system will be more complex than the current one. We can have several levels for the player to reach when his/her score is high enough. One could also get some new abilities when playing for a long time or reach a specific level, for example, clearing all the fruits on the screen with a special gesture, or destroy the bombs without getting hurt, in a specific timespan.

However, since the game system design is not the key point of this book, we will only use the two simplest concepts of a game here, that is, score and life. They are enough for interacting with the existing elements in the scene. If slicing a fruit is detected, the score increases. The life decreases if you slice a bomb. If the life is zero, the game is over and you can only restart it. The score and life values will be reset.

Additional information

Now it depends on you to improve this game as much as possible. You can change the images it currently uses. You can also add new game concepts such as levels and special props. As well as this, you can add audio and networking supports as you wish. After all, it's a Kinect-based game, which we have developed from zero!

Summary

In this last chapter, we have integrated all the past functionalities, built a simple Fruit Ninja game, and added some game elements to make it playable. You will see that with the knowledge we have collected in the last few chapters, it is not too difficult for us to develop a complete Kinect-driven game. The game itself can have more features such as audio, graphics, and networking, but this is beyond the scope of this book. You can play with your imagination now to use Kinect and its APIs and to develop more interesting and complicated applications from now on.

Where to Go from Here

Congratulations! We have already finished all six chapters of this book. We have learned how to install and configure the Kinect SDK, how to initialize it and obtain image-streaming data into OpenGL textures, how to get a player skeleton and use hand positions to emulate multitouch inputs and gestures, and how to design a comparatively complete Fruit Ninja game using these features. Now you can feel free to make use of the mighty Kinect and its APIs to develop your own applications.

But before that, in this last chapter we will provide some extra ideas about Kinect programming. First, we will have a quick look at two third-party SDKs that can replace the functionalities of the Microsoft Kinect SDK. For non-Windows developers, they are always preferred as alternative middleware. Then, we will introduce some open source and commercial software based on Kinect, and even some hardware solutions that can be considered for motion-sensing uses.

libfreenect – the pioneer of Kinect middleware

The Kinect sensor was first launched for Xbox 360 game consoles in November 2010, but the first distribution of the Kinect SDK for Windows was released in June 2011. During the interregnum, many hackers and programmers had published their methods to drive and use Kinect features, which were extremely attractive to new media artists and developers. So here come the libfreenect and OpenNI libraries.

libfreenect was born in the race of hacking Microsoft Kinect in early November 2010. Héctor Martin made his code open source on GitHub, a famous social coding host. And this is the rudiment of libfreenect. Now this library is maintained by the OpenKinect community; it can be downloaded at:

- **The community page**: `http://openkinect.org/wiki/Main_Page`
- **The GitHub source repository**: `https://github.com/OpenKinect/libfreenect`

The libfreenect library can be used with Linux, Mac OS X, and Windows. Currently, it supports RGB and depth images through the Kinect USB camera. It has a planning analyses library that provides skeleton tracking, hand tracking, audio, point cloud, and 3D reconstruction features. But it has still not been released at the time of writing. libfreenect also supports different programming languages besides C/C++, including C#, Java, Python, Ruby, and ActionScript.

As an open source project, you can contribute to libfreenect at any time under the license of Apache 2.0 or GPL 2.1 (optional).

OpenNI – a complete and powerful Kinect middleware

The **OpenNI (Open Natural Interaction)** organization was also created in November 2010. It focuses on natural and organic user interfaces and develops its own framework for Kinect devices and uses. In December 2010, PrimeSense, one of the OpenNI members, released its open source drivers and motion-tracking middleware called NITE for Microsoft Kinect.

The OpenNI framework provides a series of APIs fulfilling natural interaction requirements, such as voice recognition, motion tracking, and hand and body gestures. To directly install OpenNI packages and Kinect drivers, we can go to the following web page:

`http://www.openni.org/openni-sdk/`

Before you download and use OpenNI, you'd better remove all existing Microsoft Kinect drivers from your device manager. These are shown in the following screenshot:

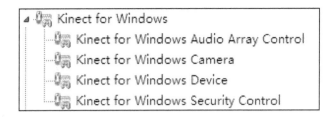

Find and remove Microsoft Kinect drivers from the Windows driver manager

You may also refer to the following article about how to install different drivers:

```
http://www.codeproject.com/Articles/148251/How-to-Successfully-
Install-Kinect-on-Windows-Open
```

Download appropriate **OpenNI Binaries** files (**stable** or **unstable**). At present, OpenNI provides Windows, Linux, and Mac OS X versions of their SDKs for use.

Install OpenNI by following the given instructions during installation. Please note that the stable version of OpenNI only supports Windows and Ubuntu at present, but its unstable version can even support Mac OS X and ARM platforms.

You may also require the NITE middleware for full-body tracking, accurate user skeleton joint tracking, and gesture recognition. It can be found at:

```
http://www.openni.org/files/nite/
```

You will find that NITE currently doesn't support ARM platforms. This means that skeleton tracking will be disabled on embedded platforms.

Now we can start an OpenNI program to see if it works. Run `NiViewer.exe` from `OpenNI\Samples\Bin\Release\`. You will see screenshots like the following one:

The result of running NiViewer

You may read something more about OpenNI programming at the following link:

`http://www.openni.org/resources/`

The OpenNI source code repository can also be found at GitHub:

`https://github.com/OpenNI/OpenNI`

PrimeSense Sensor Module for OpenNI can be found at:

`https://github.com/PrimeSense/Sensor`

`https://github.com/avin2/SensorKinect`

Free and open source resources

You will find some open source Kinect development resources and software in this section. They are both useful for learning Kinect programming in advance and making use of some valuable free resources in your own applications.

- **OpenKinect**: This can be called the wiki of the OpenKinect/libfreenect project and can be found at `http://openkinect.org/wiki/Main_Page`
- **DevelopKinect**: This is a community for Kinect depth sensor development and programming, and can be found at `http://developkinect.com/`
- **FAAST**: This is a middleware used to facilitate the integration of full-body control using skeleton tracking, and can be found at `http://projects.ict.usc.edu/mxr/faast/`
- **Coding4Fun Kinect Toolkit**: This is the utilitiy for Kinect development using C#, and can be found at `http://c4fkinect.codeplex.com/`
- **AS3Kinect**: This allows us to use Kinect functionalities in Adobe Flash and can be found at `http://www.as3nui.com/air-kinect/`
- **TUIOKinect**: This can be used to translate Kinect hand data into multitouch cursors using the TUIO protocol, and can be found at `https://code.google.com/p/tuiokinect`
- **OSCeleton**: This is used to send skeleton data in OSC format and can be found at `https://github.com/Sensebloom/OSCeleton`
- **Kinect support for Cinder (a famous open source creative-coding library)**: This can be found at `https://github.com/cinder/Cinder-Kinect`
- **Kinect for VVVV (a hybrid graphical programming environment)**: This can be found at `http://vvvv.org/documentation/kinect`
- **Kinect for openFrameworks (another open source C++ toolkit for creative coding)**: This can be found at `https://github.com/ofTheo/ofxKinect`

Commercial products using Kinect

There are some very creative solutions and products that have already been developed using Kinect as the main interaction tool. Meanwhile, talent designers are also creating some other motion-sensing systems that can totally replace Kinect in interactive applications. Here is a very incomplete list of them:

- **KinÊtre**: This can be used to create playful 3D animations using realistic deformations of arbitrary static meshes with Kinect (by Microsoft Research), and can be found at `http://research.microsoft.com/en-us/projects/animateworld/`

- **KineMocap**: This is used with multiple Kinect devices for accurate motion capturing and can be found at `http://www.kinemocap.com/`

- **FaceShift**: This uses Kinect as a replacement of motion capture devices for facial animations, and can be found at `http://www.faceshift.com`

- **ReconstructMe**: This is a real-time, 3D reconstruction system that can be used by moving Kinect around freely to collect data, and can be found at `http://reconstructme.net/`

- **So Touch Air**: Using this, you can create impressive air presentations that can be controlled by moving your hands in the air; it can be found at `http://www.so-touch.com/?id=software&content=air-presenter#`

- **OMEK**: This provides middleware and tools for easy gesture recognition and tracking interfaces, and can be found at `http://www.omekinteractive.com/`

- **SoftKinetic**: This provides 3D gesture-control solutions for consumer and professional markets, including its own time-of-flight, depth-sensing camera and the iisu SDK, and can be found at `http://www.softkinetic.com/`

- **PlayStation Eye**: This is a digital camera device for PlayStation 3; it includes computer vision and gesture recognition features and can be found at `http://us.playstation.com/ps3/accessories/playstation-eye-camera-ps3.html`

- **Xtion Pro**: This is another 3D depth camera from the OpenNI community and can currently be purchased through ASUS at `http://www.asus.com/Multimedia/Xtion_PRO_LIVE/`

- **LeapMotion**: This can be used to record and use natural hand and finger movements to precisely interact with different devices, and can be found at `https://leapmotion.com/`

Index

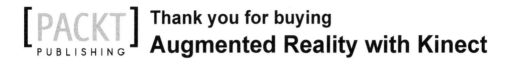
Thank you for buying
Augmented Reality with Kinect

About Packt Publishing

Packt, pronounced 'packed', published its first book *"Mastering phpMyAdmin for Effective MySQL Management"* in April 2004 and subsequently continued to specialize in publishing highly focused books on specific technologies and solutions.

Our books and publications share the experiences of your fellow IT professionals in adapting and customizing today's systems, applications, and frameworks. Our solution based books give you the knowledge and power to customize the software and technologies you're using to get the job done. Packt books are more specific and less general than the IT books you have seen in the past. Our unique business model allows us to bring you more focused information, giving you more of what you need to know, and less of what you don't.

Packt is a modern, yet unique publishing company, which focuses on producing quality, cutting-edge books for communities of developers, administrators, and newbies alike. For more information, please visit our website: www.packtpub.com.

Writing for Packt

We welcome all inquiries from people who are interested in authoring. Book proposals should be sent to author@packtpub.com. If your book idea is still at an early stage and you would like to discuss it first before writing a formal book proposal, contact us; one of our commissioning editors will get in touch with you.

We're not just looking for published authors; if you have strong technical skills but no writing experience, our experienced editors can help you develop a writing career, or simply get some additional reward for your expertise.

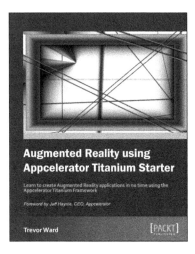

Augmented Reality using Appcelerator Titanium Starter [Instant]

ISBN: 978-1-849693-90-5   Paperback: 52 pages

Learn to create Augmented Reality applications in no time using the Appcelerator Titanium Framework

1. Learn something new in an Instant! A short, fast, focused guide delivering immediate results.

2. Create an open source Augmented Reality Titanium application

3. Build an effective display of multiple points of interest

Kinect in Motion – Audio and Visual Tracking by Example

ISBN: 978-1-849697-18-7   Paperback: 112 pages

A fast-paced, practical guide including examples, clear instructions, and details for building your own multimodal user interface

1. Step-by-step examples on how to master the essential features of Kinect technology

2. Fully-functioning code samples ready to expand and adjust to your need

3. Compact and handy reference on how to adopt a multimodal user interface in your application

Please check **www.PacktPub.com** for information on our titles

Kinect for Windows SDK Programming Guide

ISBN: 978-1-849692-38-0   Paperback: 392 pages

Build motion-sensing applications with Microsoft's Kinect for Windows SDK quickly and easily

1. Building application using Kinect for Windows SDK.

2. A detailed discussion of all the APIs involved and the explanations of their usage in detail

3. Procedures for developing motion-sensing applications and also methods used to enable speech recognition

Mastering openFrameworks: Creative Coding Demystified

ISBN: 978-1-849518-04-8   Paperback: 300 pages

Boost your creativity and develop highly-interactive projects for art 3D, graphics, computer vision and more, with this comprehensive tutorial

1. A step-by-step practical tutorial that explains openFrameworks through easy to understand examples

2. Makes use of next generation technologies and techniques in your projects involving OpenCV, Microsoft Kinect, and so on

3. Sample codes and detailed insights into the projects, all using object oriented programming

Please check **www.PacktPub.com** for information on our titles

www.ingramcontent.com/pod-product-compliance
Lightning Source LLC
LaVergne TN
LVHW080100070326
832902LV00014B/2335